Computational Social Sciences

Computational Social Sciences

A series of authored and edited monographs that utilize quantitative and computational methods to model, analyze and interpret large-scale social phenomena. Titles within the series contain methods and practices that test and develop theories of complex social processes through bottom-up modeling of social interactions. Of particular interest is the study of the co-evolution of modern communication technology and social behavior and norms, in connection with emerging issues such as trust, risk, security and privacy in novel socio-technical environments.

Computational Social Sciences is explicitly transdisciplinary: quantitative methods from fields such as dynamical systems, artificial intelligence, network theory, agent based modeling, and statistical mechanics are invoked and combined with state-of-the-art mining and analysis of large data sets to help us understand social agents, their interactions on and offline, and the effect of these interactions at the macro level. Topics include, but are not limited to social networks and media, dynamics of opinions, cultures and conflicts, socio-technical co-evolution and social psychology. Computational Social Sciences will also publish monographs and selected edited contributions from specialized conferences and workshops specifically aimed at communicating new findings to a large transdisciplinary audience. A fundamental goal of the series is to provide a single forum within which commonalities and differences in the workings of this field may be discerned, hence leading to deeper insight and understanding.

More information about this series at http://www.springer.com/series/11784

Urszula Strawinska-Zanko • Larry S. Liebovitch
Editors

Mathematical Modeling of Social Relationships

What Mathematics Can Tell Us About People

Springer

Editors
Urszula Strawinska-Zanko
Department of Conflict Resolution Studies
College of Arts, Humanities, and Social Sciences
Nova Southeastern University
Fort Lauderdale, FL, USA

Larry S. Liebovitch
Departments of Physics and Psychology
Queens College
City University of New York
Flushing, NY, USA

Physics Program
The Graduate Center
City University of New York
New York, NY, USA

Adjunct Senior Research Scholar
Advanced Consortium for Cooperation, Conflict, and Complexity (AC4)
The Earth Institute at Columbia University
New York, NY, USA

ISSN 2509-9574 ISSN 2509-9582 (electronic)
Computational Social Sciences
ISBN 978-3-030-09577-2 ISBN 978-3-319-76765-9 (eBook)
https://doi.org/10.1007/978-3-319-76765-9

Printed on acid-free paper

This Springer imprint is published by the registered company Springer International Publishing AG, part of Springer Nature.
The registered company address is: Gewerbestrasse 11, 6330 Cham, Switzerland

Acknowledgements

We would like to thank Kathleen Watkins-Richardson for valuable help in preparation of this volume.

Contents

Contributors

Andrew Z. Baker The Alliance Lab, Department of Counselor Education, Florida Atlantic University, Boca Raton, FL, USA

Clifford T. Brown Florida Atlantic University, Boca Raton, FL, USA

Nicholas Burgess Queens College, City University of New York, New York, NY, USA

Columbia University, New York, NY, USA

Peter T. Coleman Morton Deutsch International Center for Cooperation and Conflict Resolution (MD-ICCCR), Institute for Psychological Science and Practice (IPSP), Teachers College, Columbia University, New York, NY, USA

Advanced Consortium on Cooperation, Conflict and Complexity (AC4), The Earth Institute at Columbia University, New York, NY, USA

Patricia Diaz The Alliance Lab, Department of Counselor Education, Florida Atlantic University, Boca Raton, FL, USA

David Futran Queens College, City University of New York, New York, NY, USA

John M. Gottman The Gottman Institute, Seattle, WA, USA

Devin Lee Queens College, City University of New York, New York, NY, USA

Tamar Lichter Queens College, City University of New York, New York, NY, USA

Larry S. Liebovitch Departments of Physics and Psychology, Queens College, City University of New York, Flushing, NY, USA

Physics Program, The Graduate Center, City University of New York, New York, NY, USA

Adjunct Senior Research Scholar, Advanced Consortium for Cooperation, Conflict, and Complexity (AC4) at The Earth Institute, Columbia University, New York, NY, USA

Daniel Maksumov Queens College, City University of New York, New York, NY, USA

Jay L. Michaels Department of Psychology, University of South Florida Sarasota-Manatee, Sarasota, FL, USA

Jennifer P. Peluso Department of Psychology, Keiser University, Fort Lauderdale, FL, USA

Paul R. Peluso Department of Counselor Education, Florida Atlantic University, Boca Raton, FL, USA

Celine C. Ripla Northeastern University, College of Engineering, Boston, MA, USA

Ashley Sauer The Alliance Lab, Department of Counselor Education, Florida Atlantic University, Boca Raton, FL, USA

Urszula Strawinska-Zanko Department of Conflict Resolution Studies College of Arts, Humanities, and Social Sciences, Nova Southeastern University, Fort Lauderdale, FL, USA

Gabriella Tabib The Alliance Lab, Department of Counselor Education, Florida Atlantic University, Boca Raton, FL, USA

(Rae) Yunzi Tan School of Public and International Affairs, College of Public Affairs, University of Baltimore, Baltimore, MD, USA

April Watson Lynn University, Boca Raton, FL, USA

Chapter 1
Introduction to the Mathematical Modeling of Social Relationships

Urszula Strawinska-Zanko and Larry S. Liebovitch

1.1 Mathematics: From Physics to Biology to Social Science

Mathematics is logically rigorous, explicit, and focused on discovering the properties that arise from the simplest axioms assumed for real or fantasized worlds. There is an unforgiving exactness in the clarity of its statements and the logically necessary conclusions that arise from them.

Human beings are just the opposite of all this. They are emotional, complex, and implicitly sculpt and are sculpted by the rich, ever-changing, and sometimes unpredictable, multifaceted worlds that swirl around outside of them and inside of them as well. So, what can mathematics possibly tell us about social relationships?

We can best appreciate the new role that mathematics is playing in understanding social relationships by tracing the history of the application of mathematics from the physical sciences, to the life sciences, and now to the social sciences.

Over the last three centuries, mathematics has been the language of physics and chemistry. Mathematics is how these sciences express their concepts in a clear way, how the theories derived from those concepts are developed, and how the logically

U. Strawinska-Zanko (✉)
Department of Conflict Resolution Studies, College of Arts, Humanities, and Social Sciences, Nova Southeastern University, Fort Lauderdale, FL, USA
e-mail: uzanko@nova.edu

L. S. Liebovitch
Departments of Physics and Psychology, Queens College, City University of New York, Flushing, NY, USA

Physics Program, The Graduate Center, City University of New York, New York, NY, USA

Adjunct Senior Research Scholar, Advanced Consortium for Cooperation, Conflict, and Complexity (AC4), The Earth Institute at Columbia University, New York, NY, USA

U. Strawinska-Zanko, L. S. Liebovitch (eds.), *Mathematical Modeling of Social Relationships*, Computational Social Sciences,
https://doi.org/10.1007/978-3-319-76765-9_1

necessary conclusions drawn from those theories are used to make predictions. Mathematics is so intimately woven into the fabric of these sciences that Wigner (1960) could write about *The Unreasonable Effectiveness of Mathematics in the Natural Sciences*. From this approach has come our understanding of the forces of nature: gravity, electricity-magnetism, and the nuclear forces. We have used that understanding to determine the structure and function of the things in the material world around us. This understanding has also helped us to design and build buildings, bridges, airplanes, automobiles, motorcycles, bicycles, power plants, microwave ovens, computers, telephones, cell phones, satellites, communication networks, the Internet, indoor plumbing, medical imaging devices, and (for better or worse) our engines of war.

Over the last few decades, the forefront of the application of mathematics in science has subtly shifted from physics and chemistry to biology. Mathematics is now demonstrating its value to the life sciences in four different ways:

- First, as in physics and chemistry, mathematical models can determine the logically necessary consequences of specific initial assumptions, typically those of the physical laws. For example, mathematical models of the physical interactions between atoms have been used to compute the three-dimensional shapes and motions in proteins to figure out which smaller molecules, called ligands, would bind to the protein (McCammon and Harvey 1987).
- Second, mathematical analysis can "reverse engineer" a biological system to make clear the roles of its constituent units. For example, ordinary differential equation models have shown that in building a protein in the ribosome, the two-step process of transferring amino acids from t-RNA provides an error-checking mechanism that increases the fidelity of translating the information from genes to proteins. Similar methods have also shown that the multiple binding sites needed to activate a signaling protein reduce the likelihood of accidental false signals (Alon 2006).
- Third, mathematical methods can define system-level properties and provide operational definitions to measure them from experimental data. For example, the properties of networks are particularly important in understanding gene regulation, biochemical pathways in signaling and metabolism, and the connection patterns of neurons in the brain. The mathematics of graph theory and networks has introduced new concepts to biology, such as degree distribution (how many things are connected to how many other things) and motifs (types of connections, such as "feed-forward loops," where the connection between two nodes is regulated by a third node) (Alon 2006).
- Fourth, the newest methods of data science, popularly known as "big data," can analyze very much larger amounts of data than previously used methods. Just as revolutionary as the amount of data is that these methods invert the traditional perspective of analysis, replacing a top-down approach with a bottom-up approach. The traditional mathematics of physics and chemistry is top-down. It starts with assumptions and the equations that represent them and then searches for the experimental data to determine the unknown parameters in the equations.

On the other hand, these new methods are bottom-up. They start with a blank slate. It is the data itself that is used to form the associations, links, and hierarchies through methods such as "machine learning" or "deep learning." For example, this approach is being used to organize the genetic information from cancer patients and their treatments and outcomes (IBM 2016). It could also be used to design treatments with interacting multiple drugs for more targeted results and fewer side effects (Liebovitch et al. 2007).

In summary, these methods are equations based on simple assumptions to understand the functioning of an isolated piece of a system, sets of equations to reverse engineer the essential roles of the components in a system, network analysis of complex, interconnected systems, and machine learning to let the data organize itself. These four ways of applying mathematics to biology are opening up new vistas of understanding and applications in the health sciences.

These methods have proved useful in biology because they provide new and better ways to understand a complex, interconnected, multilayered, ever-changing, and sometimes unpredictable biological world. In social science we also face a social world that is complex, interconnected, multilayered, ever-changing, and sometimes unpredictable. This strongly suggests that these mathematical methods could also tell us new things about the properties of social systems.

Over the last three centuries, mathematics has given fresh insights, first to the sciences of physics and chemistry and more recently to the life sciences of biology and medicine. The next frontier for mathematical methods is their application to social science. Now is the time to use these methods to better understand social systems, the properties of the behavior of people, and the nature of social relationships. The goal of this book is to present examples of how this can be done, but even more importantly to put on clear display the shift of culture in how to think mathematically about social relationships and use that perspective to provide new insights about them.

1.2 Mathematical Models: The Good News and the Bad News

First things first. What can mathematics do for us and what can it not do for us? The following cogent passage, complete and unedited, is from Richardson (1960), who did seminal research on predicting the weather, turbulence, and fractals and whose moral values as a Quaker led him to serve as an ambulance driver in the Great War (World War I):

> To have to translate one's verbal statements into mathematical formulae compels one to carefully scrutinize the ideas therein expressed. Next the possession of formulae makes it much easier to deduce the consequences. In this way, absurd implications, which might have passed unnoticed in a verbal statement, are brought clearly into view and stimulate one to amend the formula. An additional advantage of a mathematical mode of expression is its brevity, which greatly diminishes the labor of memorizing the idea expressed. If the state-

ments of an individual become the subject of a controversy, this definiteness and brevity lead to a speeding up of discussions over disputable points, so that obscurities can be cleared away, errors refuted, and truth found and expressed more quickly than could have been done had a more cumbrous method of expression been pursued.

Mathematical expressions have, however, their special tendencies to pervert thought: the definiteness may be spurious, existing in the equations but not in the phenomena to be described; and the brevity may be due to the omission of the more important things, simply because they cannot be mathematized. Against these faults we must constantly be on our guard. It will probably be impossible to avoid them entirely, and so they ought to be realized and admitted. (xvii-xviii)

1.3 Methods

There are many different mathematical approaches to understanding social relationships described in the following chapters. We now provide an overview from 30,000 feet of that landscape to give you a sense of the many and different approaches. This is an instructive sample, rather than a comprehensive list. Additional models are also described (qualitatively) in Chap. 4: Social Processes that Generate Fractals in Brown and Liebovitch (2010).

1.3.1 Statistics

Some social scientists hear the word "statistics" when you say the word "mathematics." Just to be clear, our use of the word "mathematics" here does not mean "statistics." Sure, statistics is mathematics, no doubt about it at all. But, how mathematics is used in the physical sciences is very different than how social scientists think about statistics. This difference may not be known to social scientists who have not had firsthand experience with the culture of physical science.

In the physical sciences, you construct your model in mathematical form from the ideas you have about what is going on in the system that you are studying. The equations you write down describe the mechanisms, specifically how this changes that. How do you know what are the mechanisms to choose that you will transform into your equations? You just make it up. Yes, we stated it just the way we meant it. You just make it up! You don't completely make all of it up, of course. You use what you already know about the system, but there will still be unknown pieces, and those, yes, you just make up. When all of your thoughts and mechanisms are transformed into equations, you use mathematics to determine the logically necessary consequences of those equations. Then you see if those results make any sense, see if they match existing experimental data, and see if they make interesting new predictions that you can test later.

Social scientists looking for statistical correlations may find this surprising. Physical scientists are looking for *mechanisms*, not *correlations*. The interpretations of the equations of the model are the mechanisms of exactly how this does that. Finding those mechanisms, whose effects match the experimental data, is the goal of physical scientists. Regression models in statistics are valuable in that they show the functional correlations between variables, but most often, those models do not explain the underlying mechanisms responsible for those correlations.

The irony here is that social scientists sometimes seem squeamish that their field is less "exact" than the "hard" sciences. In fact, the truth is the oxymoron that physical scientists use *quantitative* methods only as a way to achieve a *qualitative* understanding of the physical world. A detailed example of using quantitative physical analysis to achieve a qualitative understanding of what goes on inside your eyes is given in Liebovitch (2006).

1.3.2 Differential Equations: Ordinary, Integral, and Partial

Often in physics we want to determine how the value of x over time t is changed by some other thing y. This is represented by the differential equation, $dx/dt = y$. The time derivative, dx/dt, is the rate of change of x with time t. The most famous such equation is Newton's second law, $F = ma$, where F is force, m is mass, and the acceleration $a = d^2x/dt^2$ is the second derivative of the position in space x with respect to time t. This type of model, called an ordinary differential equation, has very broad applicability to a wide range of physical and biological problems (Alon 2006; Strogatz 1994). Since these models determine how the values of the variables evolve in time from their initial conditions, they are called "dynamical systems."

In social relationships, these dynamical systems can be used in modeling how one person influences another person. Liebovitch et al. used it to describe how the emotional state (valence) of two people changes in time. The two people could be in a conflict (Liebovitch et al. 2008) or a therapist and client in psychotherapy (Liebovitch et al. 2011; Peluso et al. 2011). The equation for each person is then $dx/dt = mx + b + f(y)$ where x is the emotional state of one person at time t, m represents the degree of memory of the previous value of x, b represents the emotional state when that person x is alone, and f(y) represents the influence of the other person y on person x. With a nonlinear influence function f(y), this relatively simple model has surprisingly interesting behaviors (Liebovitch et al. 2008). Such a model can also be extended to a much larger number of people, and Fernandez-Rosales et al. (2015) used it to study up to 2048 people interacting together. Features of some differential equation models include "attractors" that are sets of values that the variables are always drawn into and the "butterfly effect" that final states can be highly sensitive to the initial conditions. These behaviors also serve as new qualitative paradigms for social systems. Coleman et al. (Coleman 2011; Vallacher et al. 2013) used such concepts from dynamical systems to shed new light on the social relationships in intractable conflicts and how to resolve them.

If one of the variables of a person, such as their accumulated emotional baggage, depends on the sum of all its instantaneous values in time, this transforms the ordinary differential equation into an integral-differential equation which requires more sophisticated mathematical methods to solve (Murray 1989).

If the variables associated with the people are not only varying in time but are also distributed across space, then the ordinary differential equation becomes a partial differential equation (Garabedian 1964). The "partial" refers to derivatives taken with respect to time with space held constant and derivatives taken with respect to space with time held constant. Such equations appear in the study of fluids, and so the techniques that have been developed to solve for the flow of air across airplane wings provide the mathematical machinery necessary to study the interactions of people over time and space. For example, such models have been used to study the patterns of settlement and transportation routes in cities (Batty and Longley 1994).

1.3.3 Difference Equations

Continuous differential equations can be approximated by their counterparts of discrete, difference equations. The variables of the model then evolve in small, discrete steps in time, Δt, rather than continuously in time. For example, the continuous derivative dx/dt can be replaced with its discrete approximations $\Delta x/\Delta t$, where $\Delta x = x(t+\Delta t) - x(t)$. With the appropriate choice of the time steps, these discrete equations very closely approximate their continuous equations. But it is not always so simple. The deep mathematics of discrete and continuous equations are actually different. The trail of the variables with time, called the trajectories, of continuous equations can never cross; otherwise the variables would not know which way to go. But the trajectories of discrete equations are sets of disconnected points which can freely cross each other. It's rare that such subtleties alter the predictions of a model, but this is a clear warning that if you don't really understand the mathematics, you can (sometimes) get into trouble.

Gottman et al. (2005) used such difference equations to describe the social relationship between a wife and her husband. In their model, first the wife responds to her husband, then the husband responds to the wife, then the wife again responds to her husband, and their dance of alternating interactions continues.

1.3.4 Self-Organizing Critical Systems

We are used to thinking that systems will fall into and remain in their lowest energy, equilibrium state. Not true for a pile of sand. As you slowly drop sand on the top of the pile, the slope steepens until it becomes unstable, and a brief avalanche of sand cascades down part of the sandpile. These avalanches remove sand until the slope is stable, which leaves each point in the sandpile caught at that razor edge between being stable and unstable. At each moment in time and at each point in space on the sandpile, the slope of the sand is at that very edge of being *unstable*. This is the opposite of a system at its most stable equilibrium. The sandpile is called a "self-organizing critical system." It organizes itself, without outside influences, to always be poised exactly at the point between stability and instability, like water at its critical point poised between a liquid and ice (Bak 1997). These systems also generate fractals in space and time. A fractal is an object with an ever larger number of ever smaller pieces, like the ever finer branches of a tree (Liebovitch 1998). The sandpile produces an ever larger number of ever smaller avalanches. Only rarely, does a very big avalanche happen. The probability of "the big one," as correctly predicted by its fractal distribution, is much larger than would be expected from a bell curve (Brown and Liebovitch 2010). "Black swans" are not "outliers"; rather they are part of the distribution typical of these systems. These models have some of the characteristics of partial differential equations because they span both space and time; however, they are implemented in discrete and iterated steps like difference equations.

Self-organizing critical systems are a good model for any system where local stress above a threshold is then shared among its nearest neighbors and so on to their neighbors. Such models have been used to better understand physical systems such as earthquakes and forest fires. They are also useful in understanding how local social relationships generate wider structures such as the viral spread of information, traffic jams, and growth patterns in cities (Turcotte and Rundle 2002).

1.3.5 Cellular Automata

Gardner (1970) described Conway's newly created game of life. It is played out on a two-dimensional board of a grid of cells like a checkerboard. The state of each cell can be "alive" or "dead." At each step in the game, the state of each cell is determined by an updating rule that depends on the states of the eight cells in the neighborhood surrounding it. It was astounding that such a simple game produced a fantastic array of patterns of great complexity that ebbed and flowed in time. This type of model, called a "cellular automata," was further generalized to cells with a larger number of states, different types of neighborhoods, and different types of updating rules. Such simple rules could reproduce many complex patterns from those found in physical phenomena to the colorful patterns on sea shells (Wolfram 1994; Meinhardt 1995). Similar to self-organizing critical systems, cellular automata span both space and time and are implemented in discrete and iterated steps. Although there is some understanding of how the patterns depend on the updating rules and neighborhood definitions (Marr and Huett 2009), in most cases the resultant patterns can only be revealed by computer simulations.

Nowak and Vallacher (1998) used these models to study the social influence on opinions. They found that complex patterns in the geographic clustering of opinions and in the polarization of opinions could arise from simple updating rules. They described this result as "dynamical minimalism" (Nowak 2004) that some complex social patterns may arise from very simple underlying rules.

1.3.6 *Non-equilibrium Thermodynamics: Complex Adaptive Systems*

Traditional analysis, in both the physical and social sciences, has often fixated on discovering the nature of a system at equilibrium in an energy minimum, because that is when it is most stable. But the most interesting and important properties of systems often do not happen at such stable states. The fluctuations in the three-dimensional structure of a protein from its energy minima may be what allows ligand molecules to bind to the protein (McCammon and Harvey 1987). The dynamical history of the players in a game may determine where they wind up, which could be away from the expected minimum of game theory (Liebovitch et al. 2008). In fact, life itself is not an equilibrium state at an energy minimum. All of life on the earth is made possible by the energy in the light flowing from the sun. Plants and animals take that energy directly, or indirectly from each other, and transform it into structures and motion, and for us, thoughts. Prigogine (1980) emphasized that it is this flow of energy from the outside through a system that creates patterns in space and time by pushing the system far from local thermodynamic equilibrium.

Kedem and Katchalsky (Katchalsky and Curran 1965) developed mathematical tools to study systems out of equilibrium, yet still close to it. The properties of systems driven by outside energy to be far from equilibrium were developed by Haken (1983); Holland (1998) and Kelso (1995). They used different types of mathematical models including differential equations, networks, and cellular automata. These systems have truly fascinating properties. They can "self-organize" creating their own patterns. They include "complex adaptive systems," which can use energy to reconfigure themselves in response to the environment around them, as well as changing the environment around them. Recall the statement about human beings at the beginning of this chapter that people implicitly sculpt, and are sculpted by, the worlds outside of them and inside of them. The properties of these mathematical models resonate with those properties of real people. Therefore, qualitative insights gained from the mathematical properties of these models have proved useful in understanding social interactions and in developing useful strategies in practical applications such as in management and the delivery of healthcare (Stacey 1992; Zimmerman et al. 1998.)

1.3.7 *Agent-Based Modeling*

Complex patterns in a large system can arise from simple interactions between its individual units. In "agent-based modeling," these individual units are represented by agents. The agents could be people deciding to invest in the stock market or automobile drivers deciding which routes to drive in a city. Each agent has a small set of values that define its state at any one time. A small set of simple rules also describes how those values change in response to the values of the other agents. For example, each agent could decide to buy or sell their stocks based on the values of their current holdings, the average value of all other stocks in the market, and whether that average value has been rising or falling. The rules could be the same for all the agents, or there could be different types of agents with different rules. The values of the agents are then evolved in time in a computer simulation. The simulation can be used to explore how the emergent patterns and dynamics of the whole system depend on the rules of the agents or the mix of different types of agents.

This method has been helpful in understanding, for example, how individual economic decisions drive prices in the marketplace, how people in a crowded theater move toward an exit, how traffic patterns adjust to a construction delay on a highway in city, and how the sociodemographic and marriage interactions determine the kinship structures in Pakistan (Bonabeau 2002; Geller 2011).

1.3.8 Data Science: "Big Data," "Machine Learning," Artificial Neural Networks, and "Deep Learning"

The mathematical approach of physical science is top-down. You first assume what mechanisms are at work in your system. Then you translate those mechanisms into equations. Maybe you also need to add some parameters specific to your particular system. Then, you solve those equations and see if those results match other existing experimental data or make new predictions that you can test later. Data science turns this process upside down, making it bottom-up. You use lots of data; petabytes (10^{15} bytes) is now typical. The data could be where people clicked on a website, retail orders, stock prices, or locations where parking tickets were issued in New York City. For social science, the data could be Twitter feeds, Google trending searches, or location check-ins on Foursquare. Typically, there's a lot of data, and you have to deal with its volume, variety, and velocity (the 3Vs). Storing, accessing, and cleaning the data are serious issues. Special software is needed to do this, for example, Hadoop is used to store and access large amounts of data in an environment distributed across clusters of computers, non-SQL programs are used to access non-relational databases, and numerical errors, misspelled words, and missing fields such as "NA" need to be cleaned up.

Mathematical and computational methods are then used to let the data organize itself. "Machine learning" tools, such as k-nearest neighbors or support vector machines, can cluster similar data together (O'Neil and Schutt 2013), or Force Atlas 2 and Gephi (Bastian et al. 2009) can identify and segment out sub-networks of nodes (which could be people, IP addresses, or Twitter hashtags). Artificial neural networks (Amit 1989) are sets of nodes connected to each other by different weights. Data values at the output nodes represent the processing of data presented at the input nodes. The network is trained by adjusting the weights so that known inputs produce their expected output values. This is called supervised learning, but networks can also be constructed with unsupervised learning, similar to the self-organization of complex adaptive systems. Between the input and output nodes are hidden layers. In ancient times (up to 2014), there were typical one or a few hidden layers. Lately, systems with hundreds of hidden layers are being used. Artificial neural networks and other computational structures that have many iterated components or processes are capable of extracting sophisticated structures from input data. Collectively, these different techniques are now called "deep learning." For example, a video of a car moving down a road on YouTube is presented to the input nodes, and the values of the output nodes could report "automobile."

These methods provide a fresh look into many different aspects of social relationships. For example, analysis of the network of Twitter users in Baltimore, MD, shows a striking lack of connection between black and white Americans, except for a possible overlap in sports teams, which could therefore play a useful role to bridge these otherwise segregated communities (Troy 2015).

1.4 Examples

Applications of mathematics have brought significant advances in theory and methodology in all areas of social science since World War II (Wilson 2010) and arguably facilitated the evolution of the study of social issues into scientific social sciences we know today. Among the so-called Big Five traditional areas of social science (Horowitz 2006), economics was the first to embrace mathematics, with applications in anthropology, political science, sociology, psychology, and allied disciplines to follow (Newman 1956; Arrow et al. 1960).

A closer look at historical developments within each social science area reveals a common pattern with an emergence of independent fields with a mathematical focus accompanied by an establishment of specialized societies and journals. For example, in psychology, mathematical psychology, defined today broadly as a field that uses mathematical methods, formal logic, or computer simulation, emerged as an independent field of research around the mid-1950s. The *Journal of Mathematical Psychology* and the Society for Mathematical Psychology were established in 1964 and 1977, respectively, and the first edition of the *Handbook of Mathematical Psychology* was published in 1963 (see Luce et al. 1963). Interestingly, the desire to frame questions about the social issues in mathematical terms continued to spread beyond the specialized journals to become an essential ingredient of contemporary social science disciplines welcomed and appreciated by all social science journals and wider audiences.

The acceleration of the adoption of mathematical approaches in social sciences is attributed to the discovery of techniques that use mathematics in qualitative ways, as well as the development of new mathematical tools, more suitable to nonphysical sciences and especially suited to investigate complex social systems where system behavior cannot be easily predicted, such as game theory (von Neumann and Morgenstern 1945), cybernetics (Wiener 1948), information theory (Shannon et al.

1949), general systems theory (von Bertalanffy 1969), diffusion models (Coleman 1964; Granovetter 1983), and dynamical system modeling (Forrester 1973; Vallacher and Nowak 1994). Fascinating mathematical models of social relationships inspired by the above approaches investigate, for example, coordination of individuals' actions (e.g., Kelso 1995; Newtson 1994; Kulesza et al. 2015), co-regulation of emotions (e.g., Babcock et al. 2013), or synchronization of physiology (e.g., Levenson and Gottman 1985). Mathematical approaches are also used successfully to shed new light on close relationships (e.g., Gottman et al. 2002; Tesser 1980) as well as interactions by more than two individuals with a focus on, to list only a few among many application areas, structure of social relationships (e.g., Boyd 1969; Freeman 2004), social influence (Nowak et al. 1990), or peace and conflict (e.g., Coleman et al. 2007; Liebovitch et al. 2010).

Mathematical applications to social sciences have been, without a doubt, revolutionized and accelerated by the use of digital computers paving way for the emergence of computational social science (CSS), an interdisciplinary field providing a unified framework for a range of research directions inspired by the paradigms of complex adaptive systems and information processing. Representative areas integrated under the umbrella of CSS are socio-informatics, computational social simulation modeling, big data analytics, social network analysis, social complexity, and several others that emerged at the intersection of traditional social science disciplines, computer science, environmental science, and engineering sciences (see, e.g., Cioffi-Revilla 2017, Vallacher et al. 2017).

1.5 The Chapters

These chapters cover a very broad range of types of social relationships and an equally broad range of mathematical models to represent them. We hope that this provides a firm starting point for you to appreciate the value and importance of mathematical modeling in providing a new perspective and understanding of social relationships.

The illustration of applications of mathematical approaches to social relationships opens with Chap. 2 authored by John Gottman and Paul Peluso titled "Dynamical models of social interaction." In this chapter, the authors present an overview of Gottman's original dynamical systems modeling work, address some of the criticisms of those models, and outline new areas of mathematical modeling to uncover additional dynamical elements of relationships.

In Chap. 3, Gottman's approach to modeling the affective exchange between interaction partners is applied in a different social context. In the chapter titled "Quantitative video coding of therapist-client sessions," Peluso and collaborators present an empirical approach to validating a model of the relationship between a therapist and their client created to tackle important questions related to success of psychotherapy.

A more in-depth presentation of this mathematical model is offered in Chap. 4 titled "Dynamical analysis of therapist-client interactions." Here, the investigators discuss model parameters, the graphical displays of the model, and the indices of model fit, and they do so through the use of clarifying case studies.

Subsequent Chap. 5 "Modeling psychotherapy encounters: Rupture and repair" presents key steps in the iterative process of a model development and refinement through an in-depth discussion of simulation work testing an extension to the mathematical model of therapeutic relationship.

In the chapter titled "Mathematical models as tools for understanding the dynamics and cooperation and conflict," Michaels shows how quantitative understanding summarized in a mathematical model may be successfully generated from theoretical ideas, observation, and empirical evidence. This time, the models presented pertain to yet another area of interests to social scientists, namely, cooperative and competitive interpersonal relations.

Chapter 7 titled "A dynamical approach to conflict management in teams" extends the presentation of relevance and practice of mathematical modeling of social relationships from an interaction between two individuals to investigations of small teams.

In Chap. 8, "Modeling the Dynamics of Sustainable Peace," Liebovitch and coauthors share details of an application of causal loop diagrams to the study of sustainable peace.

Chapter 8 titled "Capital in the first century: the evolution of inequality in ancient Maya society" takes you on a fascinating journey to the ancient Maya to enhance our understanding of how distribution of wealth evolved over time. Specifically, it discusses an application of mathematical approach to address the issue of inequality, which has always lain at the center of many social theories.

The final application presented in Chap. 10 "Can the Nash equilibrium predict the outcomes of military battles" centers on archival data and applies game theory to predict the outcome of historical military battles.

In all the chapters, the emphasis is on those aspects of the projects that show the potential of the mathematical methods presented to other social science domains beyond the specific areas featured, for example, behavioral and organizational systems.

We believe that the chapters offer a valuable and detailed practical account of efforts that are needed to prepare social science problems for an application of mathematical approaches. We also hope this volume will help and inspire readers who may not have expert knowledge of the particular research area but are interested in applying mathematical tools. We specifically asked chapter contributors to offer only as much contextual information as needed for comprehension and appreciation of the usefulness of the mathematical apparatus featured in each chapter, with the primary goal being inspiration on how it can be subsequently used by the reader in her or his own endeavors of theoretical, research, or applied nature.

References

Alon, U. (2006). *An introduction to systems biology: Design principles of biological circuits*. Boca Raton: Chapman & Hall/CRC Mathematical and Computational Biology.
Amit, D. J. (1989). *Modeling brain function: The world of attractor neural networks*. New York: Cambridge University Press.
Arrow, K. J., Karlin, S., & Suppes, P. (Eds.). (1960). *Mathematical methods in the social sciences, 1959*. Stanford: Stanford University Press.
Babcock, J. C., Gottman, J. M., Ryan, K. D., & Gottman, J. S. (2013). A component analysis of a brief psycho-educational couples' workshop: One-year follow-up results. *Journal of Family Therapy, 35*(3), 252–280 https://doi.org/10.1111/1467-6427.12017
Bak, P. (1997). *How nature works: The science of self-organized criticality*. New York: Oxford University Press.
Bastian, M., Heymann, S., & Jacomy, M. (2009). *Gephi: An open source software for exploring and manipulating networks*. International AAAI Conference on Weblogs and Social Media. https://gephi.org/users/publications/
Batty, M., & Longley, P. (1994). *Fractal cities*. New York: Academic Press.
Bonabeau, E. (2002). Agent-based modeling: Methods and techniques for simulating human systems. *Proceedings of the National Academy of Sciences (USA), 99*(Suppl. 3), 7280–7287 https://doi.org/10.1073/pnas.082080899
Boyd, J. P. (1969). The algebra of group kinship. *Journal of Mathematical Psychology, 6*(1), 139–167 https://doi.org/10.1016/0022-2496(69)90032-7
Brown, C., & Liebovitch, L. S. (2010). *Fractal analysis: Quantitative Applications in the social sciences* (Vol. 165). Los Angeles: Sage.
Cioffi-Revilla, C. (Ed.). (2017). *Computation and social science*. Springer International Publishing.
Coleman, J. S. (1964). *Introduction to mathematical sociology*. New York: The Free Press of Glencoe.
Coleman, P. T. (2011). *The five percent: Finding solutions to seemingly impossible conflicts*. New York: Public Affairs.
Coleman, P. T., Vallacher, R. R., Nowak, A., & Bui-Wrzosinska, L. (2007). Intractable conflict as an attractor: A dynamical systems approach to conflict escalation and intractability. *American Behavioral Scientist, 50*(11), 1454–1475 https://doi.org/10.1177/0002764207302463
Fernandez-Rosales, I. Y., Liebovitch, L. S., & Guzman-Vargas, L. (2015). The dynamic consequences of cooperation and competition in smallworld networks. *PLoS One, 10*(4), e0126234 https://doi.org/10.1371/journal.pone.0126234
Forrester, J. W. (1973). *World dynamics*. Cambridge: Wright-Allen Press.
Freeman, L. C. (2004). *The development of social network analysis: A study in the sociology of science*. Vancouver: Empirical Press.
Garabedian, P. R. (1964). *Partial differential equations*. New York: Wiley.
Gardner, M. (1970). Mathematical games – the fantastic combinations of John Conway's new solitaire game "life". *Scientific American, 223*, 120–123 https://doi.org/10.1038/scientificamerican1070-120
Geller, A. (2011). Growing social structure: An empirical multiagent excursion into kinship in rural north-west frontier province. *Structure and Dynamics, 5*(1). http://escholarship.org/uc/item/4ww6x6gm
Gottman, J., Swanson, C., & Swanson, K. (2002). A general systems theory of marriage: Nonlinear difference equation modeling of marital interaction. *Personality and Social Psychology Review, 6*(4), 326–340 https://doi.org/10.1207/S15327957PSPR0604_07
Gottman, J. M., Murray, J. D., Swanson, C. C., Tyson, R., & Swanson, K. R. (2005). *The mathematics of marriage: Dynamic nonlinear models*. Cambridge, MA: MIT Press.
Granovetter, M. (1983). Threshold models of diffusion and collective behavior. *Journal of Mathematical Sociology, 9*, 165–179 https://doi.org/10.1080/0022250X.1983.9989941

Haken, H. (1983). *Synergetics, an introduction: Nonequilibrium phase transitions and self-organization in physics, chemistry, and biology* (3rd ed.). New York: Springer.

Holland, J. H. (1998). *Emergence: From chaos to order*. Cambridge, MA: Perseus Books.

Horowitz, I. L. (2006). Big five and little five: Measuring revolutions in social science. *Society, 43*(3), 9–12 https://doi.org/10.1007/BF02687589

IBM. (2016). *Watson for Oncology*. http://www.ibm.com/smarterplanet/us/en/ibmwatson/watson-oncology.html. Accessed 14 June 2016.

Katchalsky, A., & Curran, P. F. (1965). *Nonequilibrium thermodynamics in biophysics*. Cambridge, MA: Harvard University Press.

Kelso, J. A. S. (1995). *Dynamic patterns: The self-organization of brain and behavior*. Cambridge, MA: MIT Press.

Kulesza, W. M., Cisłak, A., Vallacher, R. R., Nowak, A., Czekiel, M., & Bedynska, S. (2015). The face of the chameleon: The experience of facial mimicry for the mimicker and the mimickee. *The Journal of Social Psychology, 155*(6), 590–604 https://doi.org/10.1080/00224545.2015.1032195

Levenson, R. W., & Gottman, J. M. (1985). Physiological and affective predictors of change in relationship satisfaction. *Journal of Personality and Social Psychology, 49*(1), 85–94 https://doi.org/10.1037/0022-3514.49.1.85

Liebovitch, L. S. (1998). *Fractals and chaos: Simplified for the life sciences*. New York: Oxford University Press.

Liebovitch, L. S. (2006). Why the eye is round. In J. Fischbarg (Ed.), *The biology of the eye, Advances in organ biology* (Vol. 10). New York: Elsevier. http://people.qc.cuny.edu/Faculty/Larry.Liebovitch/documents/roundeyes.pdf

Liebovitch, L. S., Tsinoremas, N., & Pandya, A. (2007). Developing combinatorial multi-component therapies (CMCT) of drugs that are more specific and have fewer side effects than traditional one drug therapies. *Nonlinear Biomed Physics, 1*, 11. https://doi.org/10.1186/1753-4631-1-11

Liebovitch, L. S., Naudot, V., Vallacher, R. R., Nowak, A., Bui-Wrzosinska, L., & Coleman, P. T. (2008). Dynamics of two-actor cooperation-competition conflict models. *Physica A, 387*, 6360–6378. https://doi.org/10.1016/j.physa.2008.07.020

Liebovitch, L. S., Vallacher, R. R., & Michaels, J. (2010). Dynamics of cooperation–competition interaction models. *Peace and Conflict, 16*(2), 175–188 https://doi.org/10.1080/10781911003691625

Liebovitch, L. S., Peluso, P. R., Norman, M. D., Su, J., & Gottman, J. M. (2011). Mathematical model of the dynamics of psychotherapy. *Cognitive Neurodynamics, 5*(3), 265–275. https://doi.org/10.1007/s11571-011-9157-x

Luce, R. D., Bush, R. R., & Eugene, G. E. (Eds.). (1963). *Handbook of mathematical psychology*. New York: Wiley.

Marr, C., & Huett, M.-T. (2009). Outer-totalistic cellular automata on graphs. *Physics Letters A, 373*, 546–549 https://doi.org/10.1016/j.physleta.2008.12.013

McCammon, J. A., & Harvey, S. C. (1987). *Dynamics of Proteins and Nucleic Acids*. New York: Cambridge University Press.

Meinhardt, H. (1995). *The algorithmic beauty of sea shells*. New York: Springer.

Murray, J. D. (1989). *Mathematical biology*. New York: Springer.

Newman, J. R. (1956). *The world of mathematics* (Vol. 4). New York: Simon and Schuster.

Newtson, D. (1994). The perception and coupling of behavior waves. In R. R. Vallacher & A. Nowak (Eds.), *Dynamical systems in social psychology* (pp. 139–167). San Diego: Academic Press.

Nowak, A. (2004). Dynamical minimalism: Why less is more in psychology. *Personality and Social Psychology Review, 8*(2), 183–192. https://doi.org/10.1207/s15327957pspr0802_12

Nowak, A., & Vallacher, R. R. (1998). *Dynamical social psychology*. New York: The Guilford Press.

Nowak, A., Szamrej, J., & Latané, B. (1990). From private attitude to public opinion: A dynamic theory of social impact. *Psychological Review, 97*(3), 362 https://doi.org/10.1037/0033-295X.97.3.362

O'Neil, R., & Schutt, R. (2013). *Doing data science: Straight talk from the frontline*. Sebastopol: O'Reilly Media.

Peluso, P. R., Liebovitch, L. S., Gottman, J. M., Norman, M. D., & Su, J. (2011). A mathematical model of psychotherapy: An investigation using dynamic non-linear equations to model the therapeutic relationship. *Psychotherapy Research, 22*(1), 40–55 https://doi.org/10.1080/10503307.2011.622314

Prigogine, I. (1980). *From being to becoming: Time and complexity in the physical sciences*. San Francisco: W. H. Freeman.

Richardson, L. F. (1960). *Arms and insecurity*. Pittsburgh: The Boxwood Press.

Shannon, C. E., Weaver, W., & Burks, A. W. (1949). *The mathematical theory of communication*. Urbana: University of Illinois Press.

Stacey, R. D. (1992). *Managing the unknowable*. San Francisco: Jossey-Bass.

Strogatz, S. H. (1994). *Nonlinear dynamics and chaos*. Cambridge, MA: Perseus Books Group.

Tesser, A. (1980). When individual dispositions and social pressure conflict: A catastrophe. *Human Relations, 33*(6), 393–407 https://doi.org/10.1177/001872678003300604

Troy, D. (2015). Mapping urban social networks as a means for understanding cities. *Bloomberg Data for Good Exchange* 2015.

Turcotte, D. L., & Rundle, J. B. (2002). Self-organizing complexity in the physical, biological, and social sciences. *Proceedings of the National Academy of Sciences (USA), 99*(Suppl. 1), 2463–2465 https://doi.org/10.1073/pnas.012579399

Vallacher, R. R., & Nowak, A. E. (1994). *Dynamical systems in social psychology*. San Diego: Academic Press.

Vallacher, R. R., Coleman, P. T., Nowak, A., Bui-Wrzosinska, L., Liebovitch, L. S., Kugler, K. G., & Bartoli, A. (2013). *Attracted to conflict: Dynamic foundations of destructive social relations*. New York: Springer.

Vallacher, R. R., Read, S. J., & Nowak, A. (2017). *Computational social psychology*. New York: Routledge.

von Bertalanffy, L. (1969). *General system theory: Foundations, development, applications*. New York: G. Braziller.

von Neumann, J., & Morgenstern, O. (1945). *The theory of games and economic behavior*. Princeton: Princeton University Press.

Wiener, N. (1948). *Cybernetics or control and communication in the animal and the machine*. New York: Wiley.

Wigner, E. P. (1960). The unreasonable effectiveness of mathematics in the natural sciences. *Communications on Pure and Applied Mathematics, 13*(1), 1–14. https://doi.org/10.1002/cpa.3160130102

Wilson, T. P. (2010). On the role of mathematics in the social sciences. *The Journal of Mathematical Sociology, 10*(3–4), 221–239 https://doi.org/10.1080/0022250X.1984.9989970

Wolfram, S. (1994). *Cellular automata and complexity: Collected papers*. Reading: Addison-Wesley Pub. Co.

Zimmerman, B., Lindberg, C., & Plsek, P. E. (1998). *Edgeware: Insights from complexity science for health care leaders*. Irving: VHA Inc..

Chapter 2
Dynamic Models of Social Interaction

John M. Gottman and Paul R. Peluso

2.1 Background and Theoretical Work

The field of couples (and family) counseling is heavily dependent on the philosophy of general systems theory as outlined by Bateson and colleagues (1956), Bateson (2000). Concepts of reciprocal determinism, homeostasis, circular causality, and equifinality are all examples of key elements of family systems therapy. Bateson's original ideas were developed in the same era as cybernetics, which was developed by Weiner (1948), and game theory which was developed by von Neumann and Morgenstern (1949). According to Gottman et al. (1999), general systems theory did not have a mathematical system developed to explain or test, and "theorists of family interaction kept these systems concepts only at the level of metaphor" (p. 3). It wasn't until von Bertalanffy (1968) published his work on *general systems theory* that the beginnings of a mathematical approach to general systems were posited using differential equations (Gottman et al. 1999). However, von Bertalanffy's equations did not specify which variables were to be measured; he simply suggested that the variable be named a "g-variable." This wasn't theory; it was metaphor. Furthermore, his equations were linear equations, which are generally not stable. von Bertalanffy was apparently unaware of the nineteenth-century work of the polymath Henri Poincarè on nonlinear differential equations, which Poincarè initially used to model a solution to the three-body problem in solar gravitation.

Other than the possibility of stability, an advantage of nonlinear equations is that by using nonlinear terms in the differential equations of change, some very complex processes can be represented with very few parameters. Unfortunately, unlike many

J. M. Gottman
The Gottman Institute, Seattle, WA, USA

P. R. Peluso (✉)
Department of Counselor Education, Florida Atlantic University, Boca Raton, FL, USA
e-mail: ppeluso@fau.edu

U. Strawinska-Zanko, L. S. Liebovitch (eds.), *Mathematical Modeling of Social Relationships*, Computational Social Sciences,
https://doi.org/10.1007/978-3-319-76765-9_2

linear equations, these nonlinear equations are generally not solvable in closed functional mathematical form; solutions must be estimated numerically. For this reason, the methods are often called "qualitative," and visual graphical methods and numerical approximation are generally relied on. For this purpose, numerical and graphical methods have been developed such as *phase-space plots.* These visual approaches to mathematical modeling can be very appealing for engaging the intuition of a scientist working in a field that has no mathematically stated theory. If the scientist has an intuitive familiarity with the data of the field, our approach may suggest a way of building theory by using mathematics in an initially qualitative manner. The use of these graphical solutions to nonlinear differential equations makes it possible to talk about "qualitative" mathematical modeling. In qualitative mathematical modeling, one searches for solutions that have similarly shaped phase-space plots, which provide a good qualitative description of the solution and how it varies with the parameters (Gottman et al. 2002).

Our modeling of couples interaction using the mathematical methods of nonlinear difference equations is an attempt to integrate the mathematical insights of von Bertalanffy with the general systems theorists of family systems (Bateson et al. 1956) using nonlinear equations. In modeling marital interaction, Cook et al. (1995) developed an approach that used both the data and the mathematics of differential and difference equations in conjunction with the creation of qualitative mathematical representations of the forms of change. Our approach was unique because the modeling itself generated the equations, and the objective of our mathematical modeling was to generate theory. We suggested that the data be used to guide the scientific intuition so that equations of change were theoretically meaningful (Gottman et al. 2002).

We first needed to find a practical example of von Bertalanffy's "g-variable"; we used a report by Gottman and Levenson (1992) that one variable descriptive of specific interaction patterns is the balance between negativity and positivity because that variable was predictive of marital dissolution or stability. The challenge was also to identify what precisely was meant by "homeostasis," that is, what variable was either regulated or dysregulated in couples and families. This approach is similar to the biological approach, for example, in blood pressure regulation. We know that a blood pressure of 180 systolic over 110 diastolic puts a person at high risk for stroke, but a blood pressure of 120 systolic over 80 diastolic is healthy.

Our longitudinal research suggested that the key variable was the ratio of positive to negative affect as a couple talked to one another, especially during a conflict topic. A positive-to-negative ratio during conflict of 5.0 is healthy (predictive of relationship stability and happiness), whereas a ratio less that 1.0 is unhealthy (predictive of relationship dissolution and unhappiness). Gottman and Levenson initially used a methodology for obtaining synchronized physiological, behavioral, and self-report data in a sample of 73 couples who were followed longitudinally between 1983 and 1987. Applying observational coding of interactive behavior using Gottman's Specific Affect Coding System (abbreviated SPAFF), they computed, for each conversational turn, the number of positive minus negative speaker codes and plotted the cumulative running total for each spouse

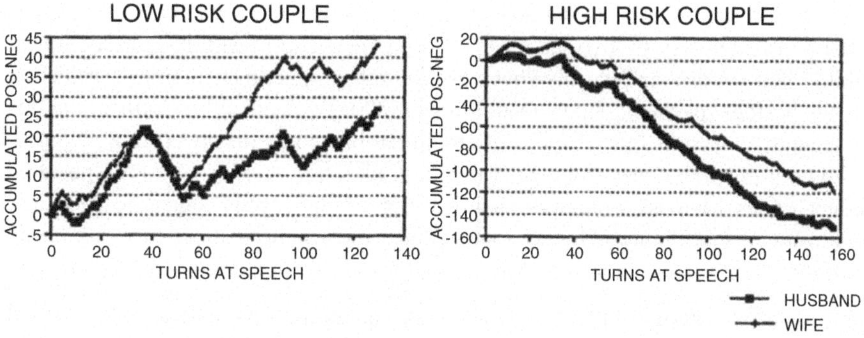

Fig. 2.1 Cumulative rapid couples interacting scoring system

(Gottman et al. 1996). As Fig. 2.1 illustrates, low-risk couples have cumulative plots that slope dramatically upward, whereas high-risk couples have cumulative plots that slope dramatically downward. The differential equation modeling used the un-cumulated data.

SPAFF Coding System Gottman developed the Specific Affect Coding System (SPAFF, Gottman et al. 1996) to code the emotional and behavioral interaction of each participant in the dyad (in their case, married couples) that had been videotaped while having a discussion about a problem. According to Gottman et al. (1996), the SPAFF focuses on the affects expressed by the participants, drawing on facial expressions, vocal tones, and speech content to characterize the emotions that are displayed. SPAFF coders "categorized the affects displayed using five positive codes (interest validation, affection, humor and joy), ten negative affect codes (disgust, contempt, belligerence, domineering, anger, fear/tension, defensiveness, whining, sadness, stonewalling), and a neutral affect code" (Gottman et al. 2002, p. 179). However, SPAFF is a cultural informants' coding system, not a specific features coding system. Although specific features are also employed (coders are trained to detect specific facial expressions and voice tones), the SPAFF coder is a highly trained observer able to integrate all cues and coding flexibly within a particular culture. SPAFF also contextualizes emotion within the interactive social context. For example, a wife's anger in response to a husband's anger might be coded as reciprocated anger toward one another, or it could be coded as empathy (high validation) if she resonates with his anger at a third person or at some stressful situation. Codes are also weighted according to their ability to predict future relationship stability and happiness. The final weighted scale ranged from −24 to +24, giving equal weighting to the positive and negative codes. These weighted time series are entered into the computer program for math modeling. The rating dial data (where partners view their video and rate the positivity or negativity of the intervention using a dial) and physiological measures (e.g., heart rate, GSR, etc.) are also entered into the math model equations (see Gottman et al. 2002 for details). Parameters are fit across a grid of possible values, and using least squares, the model with the smallest error term is accepted.

Per Gottman et al.'s (1996) protocol, for reliability assessments, each video is coded in its entirety by two independent observers. The SPAFF has excellent reliability and validity and has also been used in other social psychological studies of dyads besides married couples (Van Walsum 2005). Specifically, Gottman and his associates looked at several types of emotional behaviors displayed in a dialogue, including positive codes (interest, validation, affection, humor, and joy), negative affect codes (disgust, contempt, belligerence, domineering, anger, fear/tension, defensiveness, whining, sadness, stonewalling), and a neutral affect code (see Chap. 3 of this volume for a more detailed description of the SPAFF coding system).

Quantitative Theory In creating the model equations, a theory had to be devised that described how each partner influenced the other. The influence functions in Gottman et al. (2002) were based on two different theories of social influence we devised (which they called O-Jive and bi-linear). The O-Jive theory suggested that there is a threshold below which positive and negative affect have zero influence. The bi-linear theory suggested that at all levels positive and negative affect have a non-zero influence. The bi-linear theory fit the data best, probably because even low levels of negative affect (like a "Yes-but" statement) have a measurable negative influence on the partner. The bi-linear theory was that negative affect at time t was more powerful than positive affect in predicting the partner's next behavior at time $t + 1$. This was assessed by a steeper slope for negative than for positive affect in the influence functions. They called this theory the "triumph of negative over positive affect." The data at time $t + 1$ was modeled as consisting of the sum of a startup parameter value and an autocorrelation term at time t that assessed emotional inertia, plus the partner's influence at time t.

$$W_{t+1} = a + r_1 W_t + I_{HW}\left(H_t\right) \tag{2.1}$$

$$H_{t+1} = b + r_2 H_t + I_{WH}\left(W_t\right) \tag{2.2}$$

In these equations the parameters a and b represent uninfluenced startup of each partner, the parameters r_1 and r_2 represent emotional inertias of each partner, and the I_{HW} (H_t) and I_{WH} (W_t) represent the influence functions. Julian Cook noted that the parameters are easy to understand if one assumes that at some time points, there is zero or near-zero influence. For example, if one assumes that mutually neutral affect interactions exert near-zero influence, then one can estimate a, b, r_1, and r_2 for just those points. Then one can compute the shape of the influence function by simple subtraction:

$$W_{t+1} - a - r_1 W_t = I_{HW}\left(H_t\right) \tag{2.3}$$

$$H_{t+1} - b - r_2 H_t = I_{WH}\left(W_t\right) \tag{2.4}$$

We can then select the minimum error least squares best fitting influence function from our theoretical family of such functions. Our theoretical function included a term for what we called repair – after a threshold of negativity a switch bumping up the positive level of interaction – and a switch term for what we called turning toward, after a threshold of positivity also bumping up the positive level of interaction (see Gottman et al. 2002, for details).

Some algebra can suggest how this model provides very useful information. For example, we can compute the uninfluenced steady state by setting the influence functions to zero and fixing a value W^* and H^*. We then find that the uninfluenced steady state is:

$$W = a + r_1 W^*, \text{and} \tag{2.5}$$

$$H^* = b + r_2 H^*, \text{and solving for } W^* \text{ and } H^*, \text{we get} \tag{2.6}$$

$$W^* = a / (1 - r_1), \text{and} \tag{2.7}$$

$$H^* = b / (1 - r_2) \tag{2.8}$$

These equations show the form of the uninfluenced steady state. The uninfluenced steady state by itself was highly predictive of the relationship's stability and happiness many years later. A negative uninfluenced steady state characterized relationships that got worse over time. While their primary focus was on romantic couples, Gottman et al. (2002) also realized the potential of investigating the dynamics of change beyond the marital dyad. They wrote, "The model that was developed has given birth to a *theoretical language about the mechanism of change.* We did not have such a language before the model was successfully constructed. The model provides the language… suggests variables that can be targeted for change using interventions. In short, the model leads somewhere" (Gottman et al. 2002, p. 172).

Using algebra, we can also compute the nature of the *influenced* steady states to determine the couple's "null clines." Null clines are curves for which one partner's data do not change. These can be computed by setting $W_{t+1} = W_t = W^*$ and $H_{t+1} = H_t = H^*$ and solving:

$$W^* = [a + I_{HW}(H_t)] / (1 - r_1) \tag{2.9}$$

$$H^* = [b + I_{WH}(W_t)] / (1 - r_2) \tag{2.10}$$

Therefore, the null clines are precisely the influence functions, translated by a parameter (a or b) and stretched by another parameter ($1/(1-r_1)$ or $1/(1-r_2)$). These null clines are plotted in phase space, and their intersection determines the steady states of the couple system for that conversation. Now that we have the system's steady states, we need to determine the exact nature of these steady states. Are they stable or unstable? What is the nature of the system's return if it is perturbed? Answers to these questions are facilitated by rewriting the equations in differential form, which we now discuss.

Differential Equation Form of the Equations We can easily generalize the equations to differential form, rather than using difference equations. This important generalization was created by K.K. Tung (2007) in his book *Topics in Mathematical Modeling* (Chap. 10). The differential equations take the form:

$$\frac{dy}{dt} = f(x,y) = r_2(y - y_0) + I_2(x), y = \text{Wife} \tag{2.11}$$

$$\frac{dx}{dt} = g(x,y) = r_1(x - x_0) + I_1(y), x = \text{Husband} \tag{2.12}$$

One then linearizes these equations around a steady state, so that close to a steady state one obtains the equations:

$$\frac{dy}{dt} = f_x x + f_y y, \text{with the constants being the partial derivatives at that steady state} \tag{2.13}$$

$$\frac{dx}{dt} = g_x x + g_y y \tag{2.14}$$

The Jacobian matrix is the 2 × 2 matric of constants of partial derivatives evaluated at the steady state:

$$A = \begin{bmatrix} f_x & f_y \\ g_x & g_y \end{bmatrix} \tag{2.15}$$

The Jacobian matrix A can be used to rewrite the linear equations as **u'** = A **u**, where **u** is a column vector of x and y. The eigenvalues, denoted λ, of the matrix A come from solving the equation for the determinant, det[$A - \lambda I$] = 0, which yields a quadratic equation called the "characteristic equation":

$$\lambda^2 - \lambda[f_x + g_y] + [f_x g_y - f_y g_x] = 0, \text{usually rewritten as} \tag{2.16}$$

$$\lambda^2 - \lambda[\text{trace } A] + [\det A] = 0, \text{easily solved for potentially two values of } \lambda \text{ as} \tag{2.17}$$

$$\lambda = \frac{1}{2}\left[\text{trace } A \pm \sqrt{\left[(\text{trace } A)^2 - 4 \det A\right]}\right] \tag{2.18}$$

If the eigenvalues λs are distinct, the vector solution is of the form **u** = A exp. (λt). If the two roots of the characteristic equation (the λs) are equal, the solution is the vector **u** = ($A + Bt$) exp. (λt). If the real part of λ is negative, the steady state is stable.

The stability of the steady states is then computed using linearization close to the steady state, which involves the Jacobian matrix. The eigenvalues for each steady state are also computed. These eigenvalues describe the nature of the steady state. For example, it can be a source, a sink, a saddle, or solutions oscillating around the steady state (either away or toward it). The eigenvalues also estimate the strength of the attractor or repeller.

New Goal for Couples Therapy Ideally we would like a couple's negative steady state to have a small eigenvalue (be a weak attractor) and a couple's positive steady state to have a large eigenvalue (be a strong attractor). Then no matter where a couple begins (their uninfluenced startups), they will wind up near their positive attractors. That is a very nice state of affairs because the ratio of positivity to negativity will be high. This mathematical condition provides a new type of goal for couples therapy.

We can then understand more about the nature of the steady state using the eigenvalues, λ. If the determinant of A is positive and the trace of A is negative, then λ has a negative real part, and the steady state is stable. If $4[\det A] > (\text{trace } A)^2$, then obviously we get oscillations when the data are perturbed from the steady state.

2.2 Simulations and Intervention

It has not escaped our attention that the mathematical model makes it possible to simulate a couple's interaction patterns under imagined conditions. In the program Mathematica, for example, we create slider buttons for all the model parameters and vary the parameters systematically to determine the minimal intervention that will move a couple toward having a strong positive attractor and a weak negative attractor. The minimal reconfiguration of model parameters then suggests the intervention for a particular couple. We illustrate this procedure with a couple where changing the couple's startups moved them from having only one negative attractor to having only one positive attractor. Changing startup can be approached in many ways in Gottman couples therapy. In Fig. 2.2a we see an example of a couple in therapy whose weighted SPAFF emotional behavior was used to derive the phase-space portrait that shows a negative attractor, where the null clines meet, in the negative-negative quadrant. By varying only both partners' startup parameters so that they are positive instead of negative, the simulator gives us the phase-space portrait in Fig. 2.2b. The couple now has an attractor (where the null cines meet) in the positive-positive quadrant. Now the therapist can work on interventions that create positive startups in both partners. This would represent positive growth and changes for the couple toward a presumably more fulfilling relationship, if used retrospectively. Or, this type of modeling approach, if used prospectively, can guide couples therapists toward formulating treatment goals and directing the clinical focus in a way that would potentially change the underlying parameters to achieve the desired outcomes (as in Fig. 2.2b).

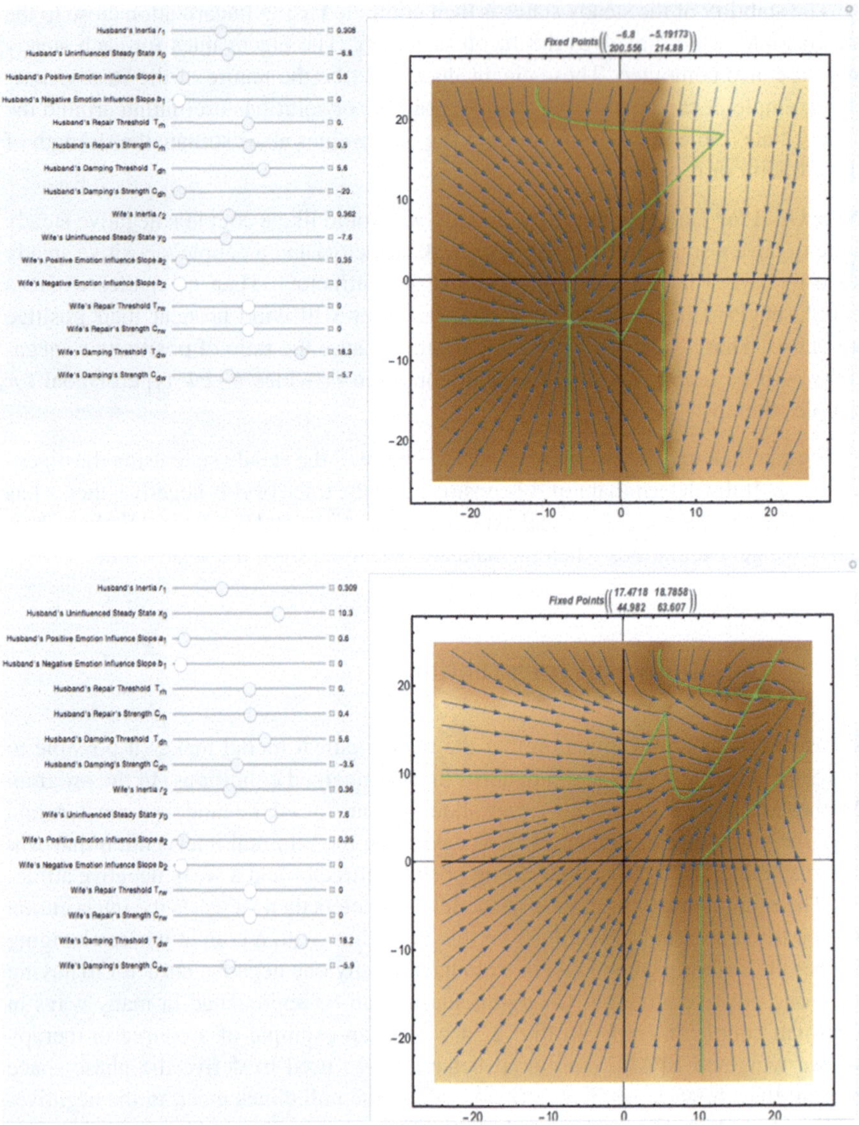

Fig. 2.2 (**a**) Couple in therapy whose weighted SPAFF emotional behavior was used to derive the phase-space portrait that shows a negative attractor. (**b**) Simulated couple where the phase-space portrait shows a positive attractor

2.3 Using Physiological Data

Any two synchronized time series (e.g., from an interacting husband and wife) can be used for mathematical modeling using the Gottman-Murray equations, assuming that there are an adequate number of data points and no missing data. We have been experimenting with phase-space plots of inter-beat interval data (IBI). Inter-beat interval is the time in microseconds between the R-spikes of the electrocardiogram. We know that once the inter-beat interval drops below 600 ms (above 100 beats a minute), it is the point at which most people start secreting adrenaline (Rowell 1993). Therefore, we can subtract 600 ms from the IBI data and create a phase-space plot. A stable steady state (attractor) in the positive-positive quadrant suggests that the couple has an attractor that pulled for physiological calm during that interaction. On the other hand, a stable steady state (attractor) in any other quadrant suggests that the couple has an attractor that pulled for physiological arousal during that interaction. Alternatively, we can subtract the mean. Figure 2.3 shows the

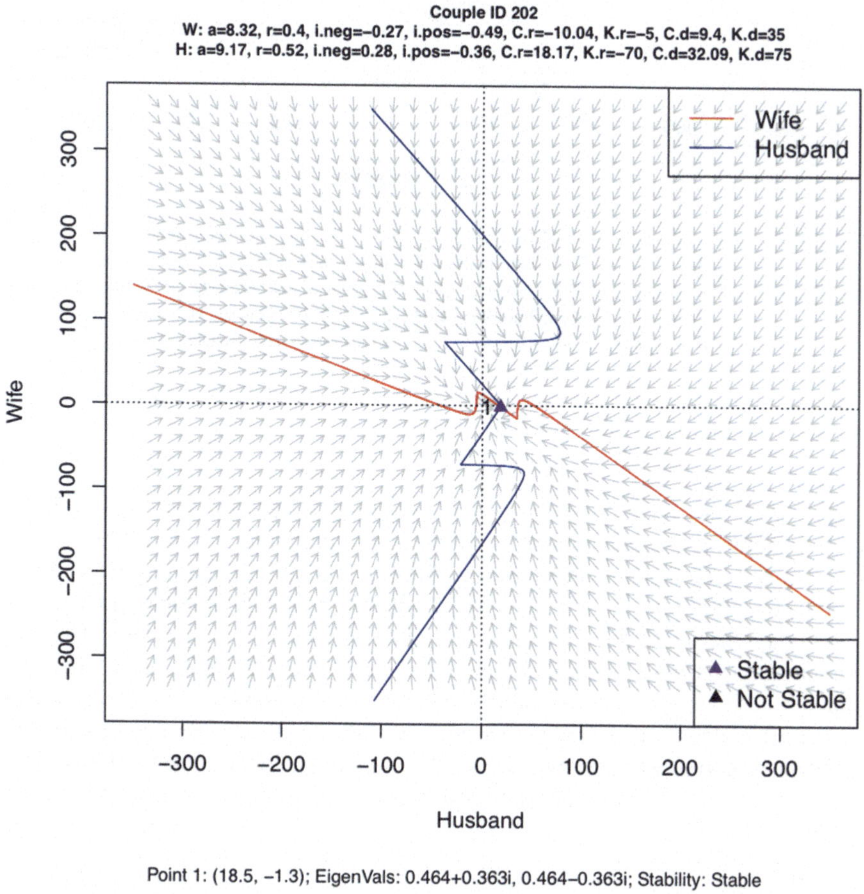

Fig. 2.3 Phase-space plot for a couple whose IBI phase-space plots show a stable point where the wife's heart rate is low, while the husband's heart rate is stable at his mean

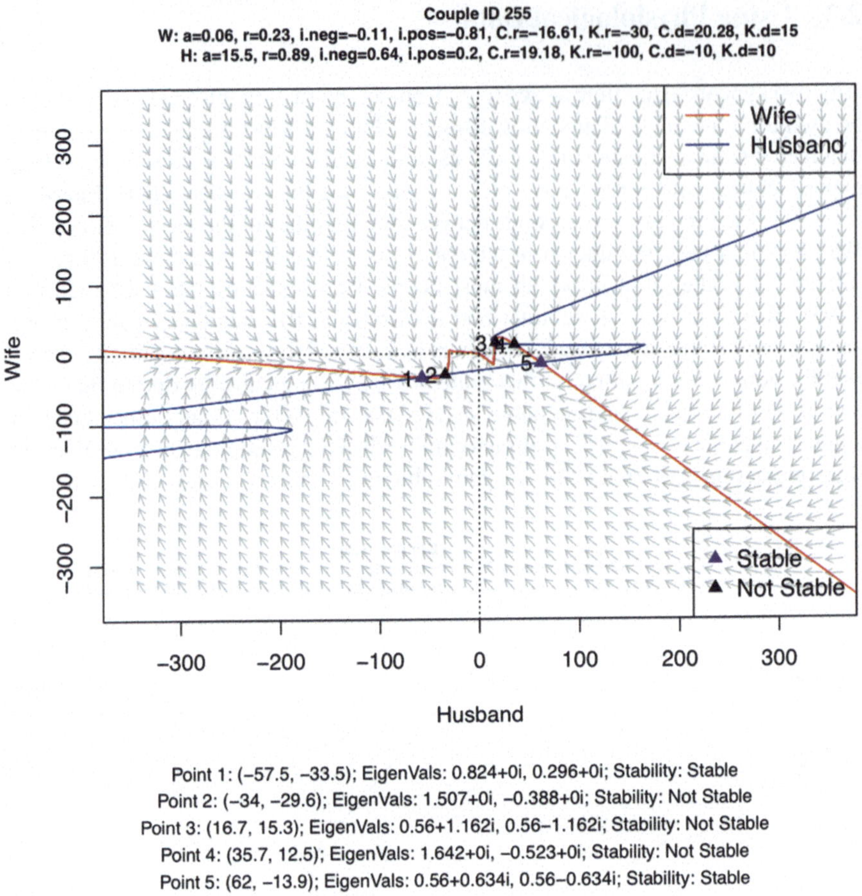

Fig. 2.4 Phase-space plot for a couple whose IBI phase-space plots show stable and unstable attractor in the negative-negative quadrant

phase-space plot for a couple whose IBI phase-space plots during their conflict conversation show a stable point where the wife's heart rate is low, while the husband's heart rate is stable at his mean. Since that couple's mean heart rates were both below 100 beats a minute, this figure suggests a positive configuration.

On the other hand, Fig. 2.4 suggests quite the opposite configuration. The two attractors in the positive-positive space are unstable. As a result, even when the couple finds themselves in this state of relative calm, it probably doesn't take much to perturb them away from it. Instead, the stable and unstable attractor in the negative-negative quadrant suggests that this is more likely the norm for this couple, meaning that there is more physiological arousal and thus a greater propensity for conflict.

2.4 Model Validation

A frequent problem that researchers using dynamical systems have faced is estimating the parameters from the data and then applying the best fit to the data. Models can be created, but if there is no way to evaluate their goodness of fit to real data, then the models are illustrative and speculative at best, but may not hold up to the most basic principle of science: falsifiability. Recently, Hamaker and her colleagues (2009) devised a method for estimating the parameters for a model and computed a Bayesian information criterion (BIC) as well as Akaike's information criteria (AIC) which can then be used to compare models using different influence functions and determine the best fit for the given interaction. Each method for assessing the goodness of fit in a model places different penalties for the number of parameters in the model, as well as different underlying assumptions about whether a "true" model exists (see Hamaker et al. for a discussion). Since the combination model has the most parameters, and the no influence model has the least, it will be important to consider both of these fit indices to insure that a well-fitting model is not rejected (type 2 error). In fact, Madhyastha et al. (2011) have recently used this to test Gottman et al.'s (2002) original influence functions. These goodness of fit tests are also a part of the R "dyad" program and are computed at the same time as the parameters for the models are estimated. Thus, employing the same approach will help to determine the best fitting model for therapists and clients, which may solve the problem of determining the effect of therapist responsiveness (differing levels of affective involvement that emerges within the context of client needs) on client outcomes (see Chap. 4 of this volume for a more thorough treatment and example of this topic).

2.5 Extension to Therapeutic Relationship

Within the last decade, Peluso, Liebovitch, and Gottman have collaborated on applying Gottman and Levenson's approach to the therapeutic relationship (Liebovitch et al. 2011; Peluso et al. 2012). Utilizing Gelso's (2009) paradigm of the "real relationship" as a key component within the therapeutic relationship as a guiding principle, we have attempted to derive the emotional qualities of successful and unsuccessful therapeutic relationships. We have employed Gottman's SPAFF and have used his dynamical systems equations to explore these dynamics from the inside out. Chapter 3 of this text details the applications of SPAFF and analysis of time series data to gain and understand the workings of the therapeutic relationship on a second-by-second basis. In Chap. 4, the mathematical modeling of the therapeutic relationship is presented, which outlines how determining the parameters and ultimately the phase-space portraits could yield further insights into the nature of successful and less successful therapeutic encounters. In addition, Chap. 5 of this text outlines how modeling repair and dampening terms in the dynamical system may be able to produce effective feedback to counselor and clients in making behavioral change.

2.6 Conclusion

Gottman's work over the last 45 years, as briefly reviewed in this chapter, has demonstrated that a dynamical systems approach can be effectively used to model one of the most complex relationships in nature: romantic relationships between two partners (REFS). Further, this work has led to the creation of an approach that can be applied beyond the human romantic dyad to other areas of human interaction. By applying this to relationships where successful outcomes are often "high stakes" or even "life and death," we feel that it may bring about some of the most meaningful and worthwhile applications of mathematics and the sciences: helping others to lead a better life.

References

Bateson, G. (2000). *Steps toward an ecology of mind*. Chicago: University of Chicago Press.

Bateson, G., Jackson, D. D., Haley, J., & Weakland, J. (1956). Toward a theory of schizophrenia. *Behavioral Science, 1*, 251–264. https://doi.org/10.1002/bs.3830010402

Cook, J., Tyson, R., White, J., Rushe, R., Gottman, J., & Murray, J. (1995). Mathematics of marital conflict: Qualitative dynamic mathematical modeling of marital interaction. *Journal of Family Psychology, 9*(2), 110–130. https://doi.org/10.1037/0893-3200.9.2.110

Gelso, C. J. (2009). The real relationship in a postmodern world: Theoretical and empirical explorations. *Psychotherapy Research, 19*(3), 253–264. https://doi.org/10.1080/10503300802389242

Gottman, J. M. (1979). *Marital interaction: Experimental investigations*. New York: Academic Press.

Gottman, J. M., & Levenson, R. W. (1992). Marital processes predictive of later dissolution: Behavior, physiology, and health. *Journal of Personality and Social Psychology, 63*, 221–233. https://doi.org/10.1037/0022-3514.63.2.221

Gottman, J. M., Coan, J. A., Carrère, S., & Swanson, C. (1998). Predicting marital happiness and stability from newlywed interactions. *Journal of Marriage and the Family, 60*, 5–22. https://doi.org/10.2307/353438

Gottman, J. M., McCoy, K., Coan, J., & Collier, H. (1996). The specific affect coding system (SPAFF). In J. M. Gottman (Ed.), *What predicts divorce? The measures*. Mahwah: Lawerence Erlbaum Associates.

Gottman, J., Swanson, C., & Murray, J. (1999). The mathematics of marital conflict: Dynamic mathematical nonlinear modeling of newlywed marital interaction. *Journal of Family Psychology, 13*(1), 3–19. https://doi.org/10.1037/0893-3200.13.1.3

Gottman, J., Murray, J., Swanson, C., Tyson, R., & Swanson, K. (2002). *The mathematics of marriage: Dynamic nonlinear models*. Cambridge, MA: MIT Press.

Hamaker, E. L., Zhang, Z., & Van der Maas, H. L. J. (2009). Using threshold autoregressive models to study dyadic interactions. *Psychometrika, 74*, 727–745. https://doi.org/10.1007/s11336-009-9113-4

Liebovitch, L. S., Peluso, P. R., Normal, M. D., Su, J., & Gottman, J. M. (2011). Mathematical model of the dynamics of psychotherapy. *Cognitive Neurodynamics, 5*, 265–275. https://doi.org/10.1007/s11571-011-9157-x

Madhyastha, T. M., Hamaker, E. L., & Gottman, J. M. (2011). Investigating spousal influence using moment to moment affect data from marital conflict. *Journal of Family Psychology, 25*(2), 292–300. https://doi.org/10.1037/a0023028

Peluso, P. R., Liebovitch, L. S., Gottman, J. M., Norman, M. D., & Su, J. (2012). A mathematical model of psychotherapy: An investigation using dynamic non-linear equations to model the therapeutic relationship. *Psychotherapy Research, 22*(1), 40–55. https://doi.org/10.1080/10503307.2011.622314

Rowell, L. (1993). *Human cardiovascular control*. New York: Oxford University Press.

Tung, K. K. (2007). *Topics in mathematical modeling*. Princeton: Princeton University Press.

Van Walsum, K. L. (2005). *Transference effects on student physicians' affective interactions and clinical inferences in interviews with standardized patients: An experimental study* (Doctoral dissertation). Retrieved from http://repository.tamu.edu/bitstream/handle/1969.1/2548/etd-tamu-2005B-CPSY-van.pdf?sequence=1

von Bertanlanffy, K. L. (1968). *General system theory: Foundations, development, applications*. New York: George Braziller.

von Neumann, J., & Morgenstern, O. (1949). *Theory of games and economic behavior*. Princeton: Princeton University Press.

Weiner, N. (1948). *Cybernetics: Or control and communication in the animal and the machine*. Cambridge, MA: MIT Press.

Chapter 3
Quantitative Video Coding of Therapist-Client Sessions

Paul R. Peluso, Patricia Diaz, and Gabriella Tabib

3.1 Introduction

The leading cause of disabilities in the United States is mental illness, and the rates of mental health disorders are staggering. According to the National Institute of Mental Health (2010), approximately 58 million people, or one in four adults in the United States, suffer with a diagnosable mental illness. Forty (40) million people are diagnosed with anxiety disorder, and approximately 14 million people are diagnosed with major depression. Psychotherapy has been shown to be effective for treating a wide array of mental illnesses. However, only 25% of individuals who have a diagnosable mental illness will come to therapy; and of those, approximately half will drop out after the first session (Mozdzierz et al. 2014; Norcross 2011). This rate of attrition has remained consistent over the last 50 years. The major question that has confronted clinicians and researchers alike continues to be: if therapy is successful and effective, then why don't clients stay?

There is substantial published evidence that the success of therapy does not strongly depend on the type of therapy (Lambert and Barley 2001; Norcross and Wampold 2011; Wampold 2010) but depends instead on the human relationship between the therapist and client. Bruce Wampold, a leading psychotherapy researcher, conducted several meta-analyses of research designed to show the superiority of one approach over another (e.g., cognitive-behavioral therapy vs. "supportive" therapy). Instead, he reported: "(i)t appears that for the most prevalent

P. R. Peluso (✉)
Department of Counselor Education, Florida Atlantic University, Boca Raton, FL, USA
e-mail: ppeluso@fau.edu

P. Diaz · G. Tabib
The Alliance Lab, Department of Counselor Education, Florida Atlantic University, Boca Raton, FL, USA

U. Strawinska-Zanko, L. S. Liebovitch (eds.), *Mathematical Modeling of Social Relationships*, Computational Social Sciences,
https://doi.org/10.1007/978-3-319-76765-9_3

disorders of adults and children, all treatments intended to be therapeutic are equally effective" (2010, p. 60). Similarly, the recent report of the American Psychological Association's task force on evidence-based therapy relationships (2011) commissioned a number of meta-analyses designed to investigate various aspects of the therapeutic relationship and therapeutic alliance. They came forth with the following conclusions.

The therapy relationship accounts for why clients improve (or fail to improve) at least as much as the particular treatment method. Practice and treatment guidelines should explicitly address therapist behaviors and qualities that promote a facilitative therapy relationship. Efforts to promulgate best practices or evidence-based practices (EBPs) without including the relationship are seriously incomplete and potentially misleading (Norcross and Wampold 2011, p. 423).

The therapeutic relationship is characterized by "the feelings and attitudes that counseling participants have towards another, and the manner in which these are expressed" (Gelso and Carter 1985, p. 159). The therapeutic alliance, a related construct, refers to the capacity of the therapist and client to form a relational bond (Horvath et al. 2011). There is much research evidence to support the importance of the role of therapeutic relationship and alliance in client outcomes. Norcross and Lambert (2011) reviewed hundreds of meta-analyses encompassing thousands of outcome studies in psychotherapy and propose a model of total psychotherapy outcome variance. The authors state that unexplained variance in psychotherapy outcomes accounts for 40% of the total variance. They also propose that the individual therapist accounts for 7%, the treatment method accounts for 8%, patient contribution (client-specific factors) accounts for 30%, and the therapeutic relationship accounts for 12% of the total variance in psychotherapy outcomes. The investigators reported that other factors account for the remaining 3%. Of the variables listed, the therapeutic relationship is of particular importance as this is the most impactful variable contributing to client outcome that can be directly modified by the therapist.

The therapeutic alliance has also been linked to the success of treatment outcomes across a broad section of clients, treatments, and identified problems (Fluckiger et al. 2012). Sharf et al. (2010) found that there was a moderately strong relationship ($d = 0.55$) between the psychotherapy dropout rate and the therapeutic alliance. Several studies have shown that the therapeutic alliance predicts symptom reduction in clients with a variety of presenting problems such as depression and eating disorders (Crits-Christoph et al. 2011; Falkenstrom et al. 2013; Gaston et al. 1991; Zuroff and Blatt 2006). The alliance also has a robust impact on outcome with a variety of treatment modalities and has been shown to be a significant predictor of client outcomes in family therapy and youth therapy, as is cohesion, a correlate of the alliance, in group therapy (Johnson et al. 2005; Norcross 2010).

However, research into understanding the qualities of the therapeutic relationship and the interaction between the therapist and client that are most effective to help change negative behaviors or habits have not had a significant breakthrough in over 30 years (Wampold 2010). By looking at the dynamical mechanisms of the therapeutic relationship in depth and in detail, we aspire to shed some light on this important area and to broaden the impact beyond the realm of psychotherapy. The potential implications for these comprehensive findings are wide-ranging and include being able to train therapists to be maximally effective with clients to change negative behaviors or habits. These findings may also generalize to other helping or health-related disciplines (e.g., medicine, nursing) where information about, advice on, and influence on positive, healthy behavior is a prerequisite for success at changing a patient's behaviors or his or her adherence to a medical regimen. However, in order to develop a better understanding of the therapeutic relationship, it is important to consider novel and underutilized approaches to this research. Peluso et al. (2012) explained, "the essentials of how to create [the therapeutic] relationship—the interaction between client and therapist—have not been modeled or observed in a direct fashion" (p. 42).

The study of the therapeutic relationship has typically occurred through pencil-and-paper measures and has frequently focused on the collaborative component of the relationship: the therapeutic alliance. Some commonly used alliance measures include the California Psychotherapy Alliance Scale (CALPAS, Gaston and Marmar 1994), the Helping Alliance Questionnaires (HAq, Alexander and Luborsky 1987), the Vanderbilt Psychotherapy Process Scale (VPPS, O'Malley et al. 1983), and the Working Alliance Inventory (WAI, Horvath and Greenberg 1986). Using measures like these, research on the therapeutic relationship is typically conducted from the perspectives of the therapist or client; however, these perspectives may be limited. Rarely is the perspective of an observer considered. One reason for the paucity of observational research in psychotherapy is that this approach is more time-consuming as well as more resource-intensive.

Results of studies examining the congruence between therapist and client scores of the therapeutic alliance suggest that only a small amount (13%) of the variability in client alliance ratings can be predicted by their therapists' assessments of the alliance (Atzil-Slonim et al. 2015). Zandberg et al. (2015) reviewed the literature on the correlations between client- and therapist-rated alliance and found them to be weak to moderate (ranging from $r = 0.07$ to 0.43) at the most (Hersoug et al. 2002; Meier and Donmall 2009; Langhoff et al. 2008). These findings are indicative of the fact that therapists' and clients' perceptions of the relationship are fairly discrepant and may not represent the most accurate appraisal of the quality of the therapeutic bond. Another way to study the therapeutic relationship is through the use of observational coding. Observational coding involves the categorical description of behavior from the perspective of a third-party observer. This approach allows for the elimination of bias related to direct participation in the therapeutic interaction. Because the therapeutic relationship is, at its core, an emotional interaction, affect coding systems are particularly helpful in measuring this construct. While there are various systems which could be used to code affect, the Specific Affect Coding System is one of the most reliable.

3.2 SPAFF Coding System

Gottman and his associates developed the Specific Affect Coding System (SPAFF, Gottman et al. 1996) in order to code the emotional and behavioral interaction of each participant in the dyad (in their case, married couples) that had been videotaped while having a discussion about a problem. According to Gottman et al. (1996), the SPAFF focuses solely on the affects expressed by the participants, drawing on facial expressions, vocal tones, and speech content to characterize the emotions that are displayed. SPAFF coders "categorized the affects displayed using five positive codes (interest, validation, affection, humor, and joy), ten negative affect codes (disgust, contempt, belligerence, domineering, anger, fear/tension, defensiveness, whining, sadness, and stonewalling), and a neutral affect code" (2002, p. 179). The final weighted scale ranged from −24 to +24, giving equal weighting to the positive and negative codes. Table 3.1 presents each of the SPAFF codes in order from negative through neutral to positive, along with a definition of each.

According to Heyman (2001), the SPAFF "has by far the best evidence of construct and criterion validity" (p. 25). At present, however, Gottman's research has primarily been applied to couples and their dynamics, though it is a method that can be used with other dyadic exchanges (Gottman et al. 2002). For example, Main (2015) successfully applied the SPAFF coding system to a sample of mother-adolescent dyads. She specifically found that empathic communication (evidenced by the codes of "interest" and "validation") by mothers in the face of negative emotion in adolescents was predictive of successful conflict management. Giese-Davis et al. (2005) used SPAFF to look at affective expressions of women in group therapy sessions for breast cancer treatment. Van Walsum (2005) expanded the method to look at doctor and patient affective exchanges. Student physician sessions were videotaped and affect coded with the modified SPAFF. The purpose was to look at the effects of transference on affective exchanges. Her results revealed that negative emotion, coded by SPAFF, had an effect on simulated patients, particularly by gender (Van Walsum 2005). Lastly, Erzar et al. (2012) have recently published a Slovenian translation of SPAFF which they applied to a counselor-client dyad. They presented several clinical vignettes that illustrated the application of SPAFF codes to a counseling relationship. These investigations demonstrate that SPAFF can be used in relationships that are not romantic in nature.

3.3 Discriminant Function Analysis and Re-weighting

While the SPAFF has excellent reliability and validity (Gottman et al. 1996) and has also been used in other social psychological studies of dyads besides married couples (Erzar et al. 2012; Giese-Davis et al. 2005; Main 2015; Van Walsam 2005), until recently, the weighting for the SPAFF codes has been calibrated only to the husband-wife dyad. Leudke et al. (2017) conducted a discriminant function

Table 3.1 SPAFF codes definitions

SPAFF affect codes (Gottman et al. 1998)
1. *Disgust* is a negative code that includes a "verbal and nonverbal rejection of some kind of noxious stimulus."
2. *Contempt* is a negative code in which there is an attempt to insult or demonstrate lack of respect toward another. This is demonstrated with sarcasm, mockery, insults, or hostile humor.
3. *Belligerence* is a negative code that is "provocative. The belligerent person disagrees with or contests whatever her partner is saying, seemingly regardless of content. This person appears to be provoking a response in her partner, as if trying to start a fight."
4. *Low domineering* is a negative code in which the person attempts to control the conversation and force compliance and agreement.
5. *High domineering* is a negative code in which the person attempts to control the conversation and force compliance and agreement.
6. *Criticism* is a negative code where someone is attacking the other's personality where there is blaming, character attacks, long lists of complaints, or indication of betrayal.
7. *Anger* is a negative code where the person sounds fed up. For this code, the combination of tone, expressions, and body gestures are all important components.
8. *Tension* is a negative code where there is an uncomfortable air or feeling present. This can be observed through fidgeting, shifting, or nervousness.
9. *Tense humor* is when there is a mutual nervous laughter that is short and uncomfortable.
10. *Defensiveness* is a negative code in which the person feels blameless and will use "yes – but" statements as well as excuses and countercriticism.
11. *Whining* is a negative code where the vocal quality is the main component. The voice will sound nasal or singsongy.
12. *Sadness* is a negative code demonstrated by a "decrease in energy and a passive, resigned countenance." There may be crying, sighing, or pouting.
13. *Stonewalling* is a negative code where the person stops responding to their partner. This can include no vocal acknowledgments or head nods, no facial movement, and little to no eye contact.
14. *Neutral* is the division between positive and negative codes. Exchanges are a matter of fact, and there are no clear physical cues. This is otherwise known as the face at rest.
15. *Interest* is a positive code where the person is genuinely concerned and inquisitive about the other's feelings or thoughts. The person is eliciting more information from their partner.
16. *Low validation* is a positive code in which one person demonstrates acceptance to the other person's viewpoint and provides back channeling such as head nods.
17. *High validation* is a positive code where there is paraphrasing, apology, or the ability to finish the other's sentence.
18. *Affection* is a positive code that is directly in sync with caring, tenderness, compliments, or empathy.
19. *Humor* is a positive code where there are shared moments of laughter with feelings of happiness. This is demonstrated through giggling, mutual or private jokes, and good-natured teasing.
20. *Surprise/joy* is a positive code where there is a "positive, happy, or emphatic reaction to some event."

analysis, following the guidance of Gottman, to re-calibrate the weightings of the affect codes. Using dropout or return as the dependent variable, they assessed whether any combination of SPAFF scores could discriminate between clients who dropped out after one session and those who would return for a fourth session. For clients, the discriminant analysis found that SPAFF scores could significantly classify clients who dropped out compared to those who returned. The effect size, using the canonical discriminant function coefficient for the discriminant function ($r = 0.565$), suggested that 32% of the variance was explained by the discriminant function. The overall "hit rate" of correct group classification of 77% was also significantly different than chance (50%) and even surpassed the 25% above chance threshold recommended by Burns and Burns (2008) for practical significance. One noteworthy element of the result was the finding that the discriminant function allowed for a correct classification of 94.4% of returning clients but only 50% of dropout clients.

A separate discriminant function analysis was conducted to determine if SPAFF scores could discriminate between counselors whose clients dropped out after one session and those who would return for a fourth session. Again, the effect size, using the canonical discriminant function coefficient for the discriminant function ($r = 0.604$), indicated that 36% of the variance was explained by the discriminant function. The overall "hit rate" of correct group classification of 77% was also significantly different than chance (50%) and also surpassed the 25% above chance threshold recommended by Burns and Burns (2008). The discriminant function was able to correctly classify 63% of counselors who had a client drop out and 86% of counselors who had a client return.

The result of these analyses was a re-weighting of the SPAFF codes using the structure matrix as a guide (Peluso 2015). This followed the process by which the original SPAFF was weighted (J. Gottman, personal communication, June 9, 2013). The weightings of affect codes were therefore changed from the original husband-wife weightings to correspond with the therapist-client relationships. The range was kept the same as the original weighting (−4 to +4), but the weight of each affect code was modified for both therapists and clients separately in order to reflect power differentials in the relationship (in the original husband-wife weightings, they were the same for each partner). This has allowed for a more dynamic investigation into the time series of the data and more reliable interpretations of the results (Luedke et al. 2017; Peluso 2015). Re-weighting the SPAFF to be specific to counselor-client relationships might intensify the impact and clarity of patterns of observed behavior when analyzed with mathematical modeling (see Chapter 4, this volume, for more details).

3.4 Case Studies

In the next phase of the project, we began to apply the results of the simulated sessions to actual therapy sessions. The sessions were broken up into 15-min segments (to prevent coder fatigue), and then the data for the therapist and client of each

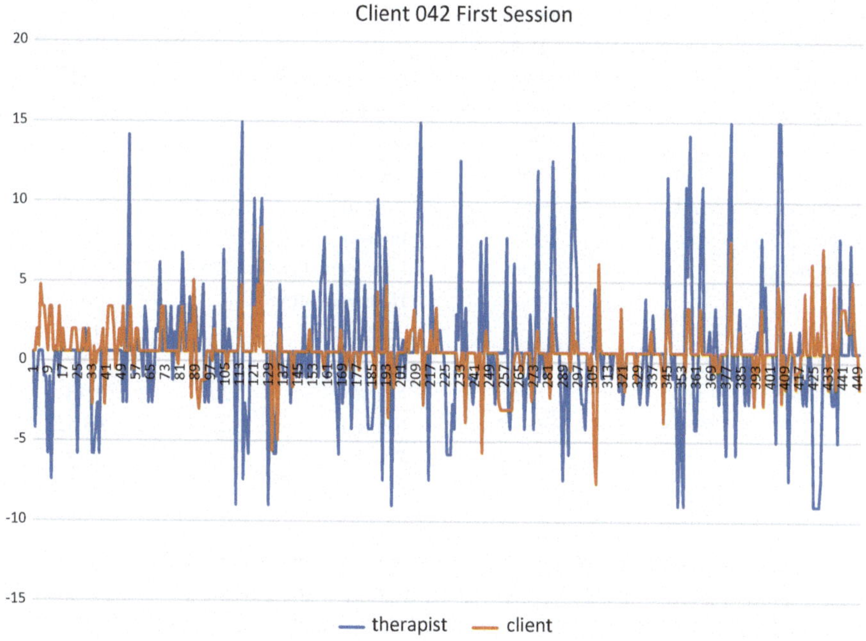

Fig. 3.1 Time series of the weighted SPAFF codes based on the first session of client number 042

session was coded, summed, and weighted in accordance with Gottman's research protocol; then it was assembled together. Time-series graphs (like the one in Fig. 3.1) were created, allowing for some interpretation of the therapist and client's emotional exchange over the time span. Levels of positive or negative affect can be seen, and areas of dynamic exchange can be identified. The most significant sessions in the sample are those that have a variety of codes as well as insightful dialogues between the therapist and client. Comparing the first and fourth sessions of such cases shows changes in the affect of the client and the working alliance alike. The following selected cases are examples of such evolution. We now present an in-depth analysis of three separate case studies.

3.4.1 *Case 042: Returned, Successful Therapy*

The first case follows a therapeutic dyad that is effective and seemingly well-connected. Table 3.2 lists the specific SPAFF codes for both the therapist and the client for this session, as well as the total number of seconds (and percentage) of positive, negative, and neutral codes. The client is visibly sad (accounting for almost 5% of all coded affect in the session), and his demeanor exudes pain and embarrassment. This behavior on the client's part is not typical for an initial session. The

Table 3.2 Number of seconds and percent of each code: Client 042 first session

Code	Therapist count	Therapist (%)	Code	Client count	Client (%)
Tension	103	3.81	Low domineering	3	0.11
Tense humor	4	0.15	Tension	94	3.48
Neutral	2027	75.07	Tense humor	5	0.19
Interest	124	4.59	Sadness	100	3.70
Low validation	261	9.67	Neutral	2336	86.52
High validation	166	6.15	Interest	2	0.07
Affection	15	0.56	Low validation	155	5.74
			High validation	3	0.11
			Affection	2	0.07
Total	2700	100		2700	100
Negative	107	3.96	Negative	202	7.48
Neutral	2027	75.07	Neutral	2336	86.52
Positive	566	20.96	Positive	162	6.00

swiftness of the client to confide in the therapist is perhaps one of the reasons their relationship appears to be strong. As for the therapist, she quickly exhibits affection. As previously mentioned, coding affection for the study's purposes meant the subject displayed mirrored affect or complimented the person on an ability or action (what is often termed "empathy"). The therapist in this case mirrors the client's sadness. When the client explains a predicament they have faced, the therapist in turn looks downcast and speaks in the same low tone when asking questions or giving insight. The therapist also allows for silence. This is indicated by the percentage of "neutral" codes (approximately 75%). When the client pauses, the therapist does not quickly fill the room with her opinions but allows for the client's words to puncture the moment. It seems to give the client a chance to truly identify with his perspectives and come to terms with saying so in front of another person.

In addition, Fig. 3.1 provides a time-series graph of the weighted SPAFF scores. Throughout the first session, the client exhibits double the rate of negative codes compared to the therapist. While the client is coded as sad and angry, the therapist only exhibits tension and tense humor. This is understandable, given that the client is divulging much about the problems that have necessitated coming to therapy. The therapist in the first session is learning about the client, so the negative spikes of tension and tense humor are quick and are followed by sharp increases in positivity, as if she is aware of her countertransference feelings but then attempts to use them constructively. Conversely, the therapist is coded as three times more positive as the client. This is seen through codes of interest, high and low validation, and affection. Interest and validation are quite common in the first session of therapy. They are generally seen in question asking, attentive listening, and rephrasing what the client has said. Affection is less common in therapy settings, but between this dyad, it seems to be effective as a way for the client to feel much more comfortable in giving background information.

Table 3.3 Number of seconds and percent of each code: Client 042 fourth session

Code	Therapist count	Therapist (%)	Code	Client count	Client (%)
Low domineering	44	1.74	Low domineering	5	0.20
Tension	89	3.52	Anger	15	0.59
Tense humor	1	0.04	Tension	89	3.52
Neutral	1684	66.51	Tense humor	2	0.08
Interest	65	2.57	Sadness	416	16.43
Low validation	144	5.69	Neutral	1957	77.29
High validation	503	19.87	Interest	1	0.04
Affection	2	0.08	Low validation	30	1.18
			High validation	14	0.55
			Affection	3	0.12
	2532			2532	
Negative	134	23.40	Negative	111	23.36
Neutral	1684	59.76	Neutral	1957	76.22
Positive	714	16.84	Positive	48	0.42

Similarly, in the fourth session of this case, the therapist and client again show affection and sadness, respectively. Table 3.3 shows the number and percentage of each SPAFF code for the therapist and client in this session. Since the client returned, it is clear that the two have created a working alliance that has fostered much growth. In the session, the client does seem more confident in his decisions, regardless of how scared and sad he seemed. However, a new and just as important code emerges during this session: low domineering. In a therapeutic context, low domineering refers to when one partner refutes the other's opinion. In some cases, the therapist may be the one exhibiting low domineering in order to offer a better means of working through problem or idea. In other cases, the client may disagree with the therapist's suggestions. In this case both happened, suggesting a "give and take" dialogue rather than a passive "one-way" discussion. It is noteworthy to state that productive and effective therapeutic relationships are not free of negative affect codes but, rather, must be presented within the context of a solid working alliance.

The therapist suggested alternatives to the client's negative way of thinking while noting that the client's current thought process would not help the client achieve their goals. The therapist was actively going against what the client believed. This can be seen in Fig. 3.2 at seconds 126 and 211. This method was well received by the client as indicated by their agreement. The therapist giving their opinion without masking it helped the client understand how to not approach their situation, which opened a dialogue for the two to talk about effective ways for the client to work through their problem. Similarly, the therapist expressed an opinion later in the session that the client outright denied. This instance of client low domineering shows the growth of the client. Who would have thought a client so tormented by sadness in the first session would grow enough by the fourth to outwardly express their opinions without fear? From the first to the fourth session, this therapeutic relation-

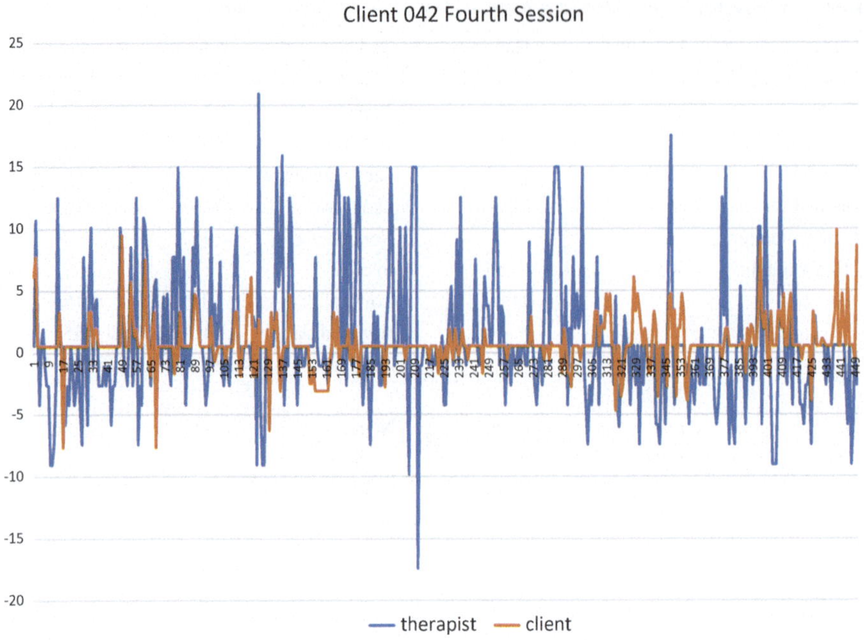

Fig. 3.2 Time series of the weighted SPAFF codes based on the fourth session of client number 042

ship grew from shared feelings to having an active conversation and planning for the future. In fact, the client admits that although he is currently experiencing difficulty at this point in his life, he felt much better after coming in for the therapy session.

3.4.2 *Case 031: Did Not Return*

A second case follows a client and therapist through a rocky first session. As this is an instance of client dropout, this session serves as a fantastic example for what both the client and therapist must bring to the relationship to foster a positive working alliance. Also, it is important to note that not all clients need more than one session. Some clients have problems that can be solved by just having someone to talk to as they piece together their thoughts out loud. Therapists in this case may need to be more passive for the client to be able to speak freely. As seen in previous cases, the therapists did not enact high percentages of negative codes until the final moments of the first session (if at all) and left negative codes mostly for the fourth session. In this session, the therapist is a much more active part of the alliance and tends to criticize more than suggest alternative behaviors for the client.

Table 3.4 Number of seconds and percent of each code: Client 031

Code	Therapist count	Therapist (%)	Code	Client count	Client (%)
Tension	221	8.19	Low domineering	2	0.07
Tense humor	13	0.48	Tension	675	25.00
Neutral	1781	65.96	Tense humor	6	0.22
Interest	48	1.78	Defensive	2	0.07
Low validation	489	18.11	Neutral	1754	64.96
High validation	146	5.41	Low validation	231	8.56
Humor	2	0.07	High validation	30	1.11
	2700			2700	
Therapist			Client		
Negative	234	8.67	Negative	685	25.37
Neutral	1781	65.96	Neutral	1754	64.96
Positive	683	25.30	Positive	261	9.67

Initially, the client seems quite apprehensive about talking to the therapist, as was seen by constant hiding behind hands and nervous laughter. The therapist starts the session by reassuring the client that being nervous is common. The therapist exhibits thoughtful validation of the client's initial statements, as seen in Table 3.4. The therapist repeats the client's observations of problems and finishes sentences 146 seconds during this session. However, the therapist visibly begins to become fatigued by the client's perpetual state of embarrassment. As seen in the high percentage of criticism, the therapist tells the client that always hiding behind their hands and stopping their train of thought will not be beneficial to working through the client's problems. The client does not seem upset by the therapist's criticism and begins to answer much more clearly and with less embarrassment.

Later in the session, the therapist asks the client what brought them to therapy. The client admits that they had made a promise to go to therapy if their friend went to therapy as well. Because of this, the client explains they do not have any goals in mind. Instead, the client felt that if they could vent their problems to their therapist, it would help them come to conclusion. It is also important to note the client's history, as it may influence why "venting" may be more helpful to them than advice-giving. The client's family is quite emotionally distant. The client also seems to be a self-appointed mediator between family members. Therefore, the client feels the need to be strong and has learned very valuable coping mechanisms on their own. The client can self-monitor and utilize online blogging to connect with others who share similar experiences. Because of the client's apathetic family and strong persona, what he truly craves is validation. The therapist and client both agree on this point. The client receives validation from his daily blog and from the therapist's active listening.

Throughout the session, the client and therapist have a relatively neutral baseline in codes as seen in Fig. 3.3. However, in Table 3.4, the percentages in overall codes are shown. It is interesting to note that the client shows 685 seconds of negative

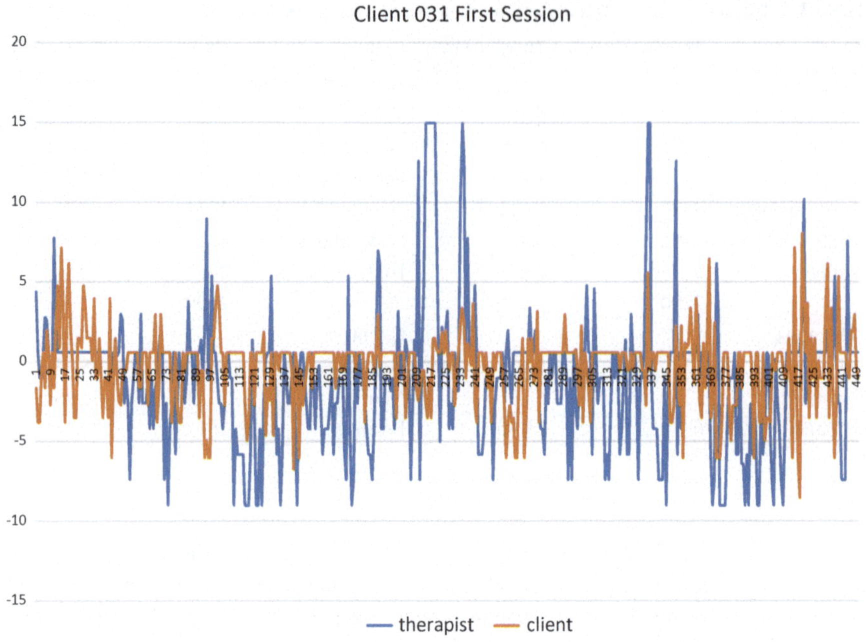

Fig. 3.3 Time series of the weighted SPAFF codes based on the first session of client number 031

codes, while the therapist only shows 234 seconds of negative affect. Conversely, the therapist shows more overall positive codes, while the client shows much less positivity throughout the session: 683 and 261 respectively. This is quite interesting mainly because when applied to the video, it shows that when the client was showing negativity, the therapist was positive. When the client exhibited positivity, the therapist was negative. This mismatch of affect may have been part of why the client did not return for the fourth session with this therapist.

In the end, the client does say that she feels much better after coming in for the session. The client admits to being "less burdened" and expresses interest in coming back for a second session. Whether or not the second session happens is unknown, but the client does not return for a fourth. Perhaps this is because their need to vent was fulfilled. Perhaps the client's problems were fixed before the fourth session. It could have also been related to the disparity in the positivity and negativity seen in the first session.

3.4.3 Case 076: Returned, Brief, Time-Limited Therapy

The next case is a perfect example of how matching personality can positively affect a working alliance. The therapist in this dyad was directive and required the client to listen and consider another point of view. Fortunately for the client, this is exactly

Table 3.5 Number of seconds and percent of each code: Client 076 first session

Code	Therapist count	Therapist (%)	Code	Client count	Client (%)
Low domineering	89	3.30	Tension	490	18.15
Tension	48	1.78	Tense humor	7	0.26
Tense humor	23	0.85	Neutral	2137	79.15
Neutral	2149	79.59	Low validation	64	2.37
Interest	124	4.59	High validation	2	0.07
Low validation	195	7.22			
High validation	72	2.67			
	2700			2700	
Negative	160	5.93	Negative	497	18.41
Neutral	2149	79.59	Neutral	2137	79.15
Positive	391	14.48	Positive	66	1.78

what was requested. The client had been in therapy many times before. Because of this, the client was aware of the therapeutic process and required a more active voice from the therapist. The client stated that she would rather have a therapist who tells them how to act instead of being guided through introspection. At the same time, the client knew that she was not willing to talk about certain aspects of themselves. For instance, when the therapist asks the client what their strengths are, the client passes and does not give in to the therapist's attempts to push the question further. This is indicated in Table 3.5 by the percent of low domineering codes in the session (over 3% of the time). In Fig. 3.4, this is tracked toward the end of the session (around the 37–42-min mark). The therapist has to gauge how far they can question the client and still keep the client in a comfortable position. This particular client must also feel that they are in control, as seen in their firm stance on not answering the question.

The time series can also be graphed cumulatively, which provides information on the overall levels of positive and negative emotion in the session. Figure 3.5 shows the same information as Fig. 3.4 but in a cumulative way over time. This presentation allows one to see the overall trajectory of the session and therefore view the relative strength or weakness of the relationship. In Fig. 3.5, the trajectory for the therapist is upward and positive for the majority of the session but then dips negative, while the client creates a sawtooth pattern on the negative side. Despite this seemingly negative outcome, the client returns.

Another interesting aspect of this case is the relatively short time period between the first and fourth session compared to other cases. The client was going to move away approximately 1 month after the first session; therefore, a brief therapy approach was needed. This meant that the goals of this therapeutic endeavor had to be clear and included creating a good foundation for the client to transition to another therapist. Part of this foundation included the therapist conducting a personality test for the client. The idea of a personality test did not seem to resonate well with the client. The client admits that they do not believe in divulging much about

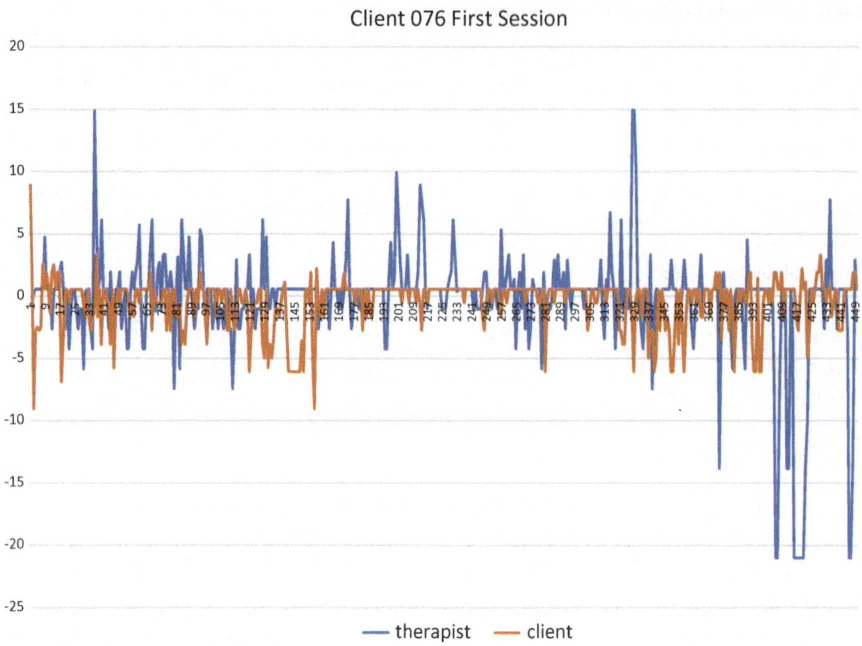

Fig. 3.4 Time series of the weighted SPAFF codes based on the first session of client number 076

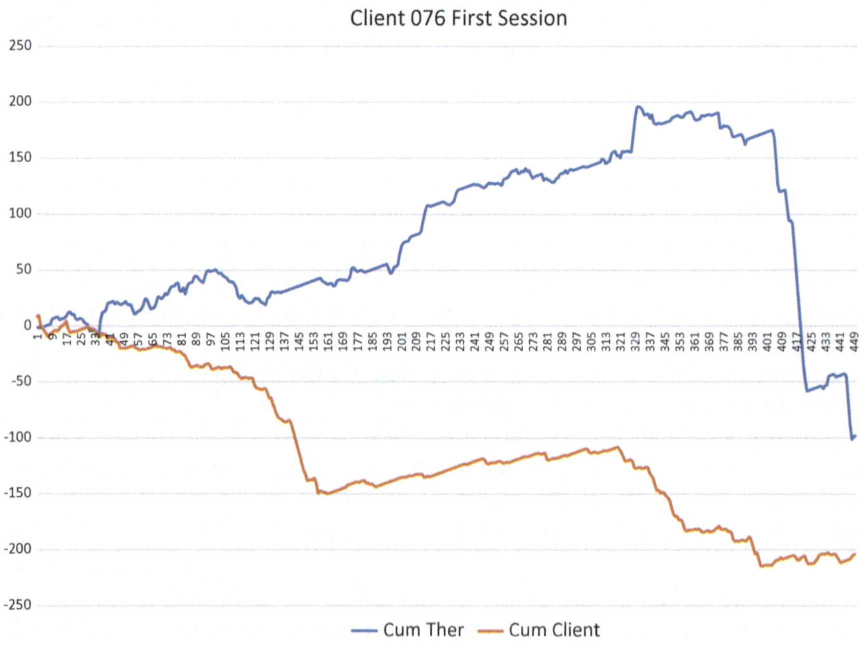

Fig. 3.5 Cumulative time series of the weighted SPAFF codes based on the first session of client number 076

Table 3.6 Number of seconds and percent of each code: Client 076 fourth session

Code	Therapist count	Therapist (%)	Code	Client count	Client (%)
Belligerence	2	0.08	Low domineering	13	0.50
Low domineering	576	21.93	Tension	744	28.33
Tension	199	7.58	Tense humor	7	0.27
Tense humor	6	0.23	Defensive	38	1.45
Defensive	3	0.11	Sadness	2	0.08
Neutral	1824	69.46	Neutral	1848	70.37
Interest	23	0.88	Interest	13	0.50
Low validation	54	2.06	Low validation	33	1.26
High validation	13	0.50	High validation	2	
	2700			2700	
Negative	786	29.11	Negative	804	29.78
Neutral	1824	67.56	Neutral	1848	68.44
Positive	90	3.33	Positive	48	1.78

themselves to other people. The therapist and client decide to table the personality test. However, it is observed in the fourth session that the two eventually did conduct the personality test, as they reference its results in the beginning of the fourth session. This implies that in the time between the first and fourth session, the client grew to trust the therapist.

Table 3.6 lists the specific SPAFF codes for both the therapist and the client for the fourth session. An examination of these codes reveals a similar activity level of both therapist and client. The therapist and client show similar counts of all codes positive, negative, and neutral. This indicates that the therapist and client are on the same page. They are mirroring each other's affect, which may have led to a better understanding of each other's perspectives.

In terms of the subject of the fourth session, the client and therapist began with a disagreement on a recent accomplishment that the client had made. The client did not consider it an achievement, but the therapist seems to think otherwise. Here, the therapist decides it is time to introduce new methods of thinking for the client. The client's comparisons to others pose a threat to their progress in therapy, and the therapist does not sugarcoat the need for the client to change them. As seen in Table 3.6, the therapist exhibits proportionally higher negative codes in the fourth session compared to the first session. Many observers may consider exhibiting negative codes to be counterproductive in creating a positive working alliance. However, it is important to note that just because a subject exhibits negative codes, it does not mean they have malicious intentions.

For instance, the therapist displays belligerence toward the client (seen on Table 3.6 as code 3). The client is adamant that being logical is the only perspective they need to navigate through life. The therapist does not validate the client's absolute point of view, and when the client becomes defensive, the therapist continues to further a deeper interpretation of what it means to always be logical. Although being

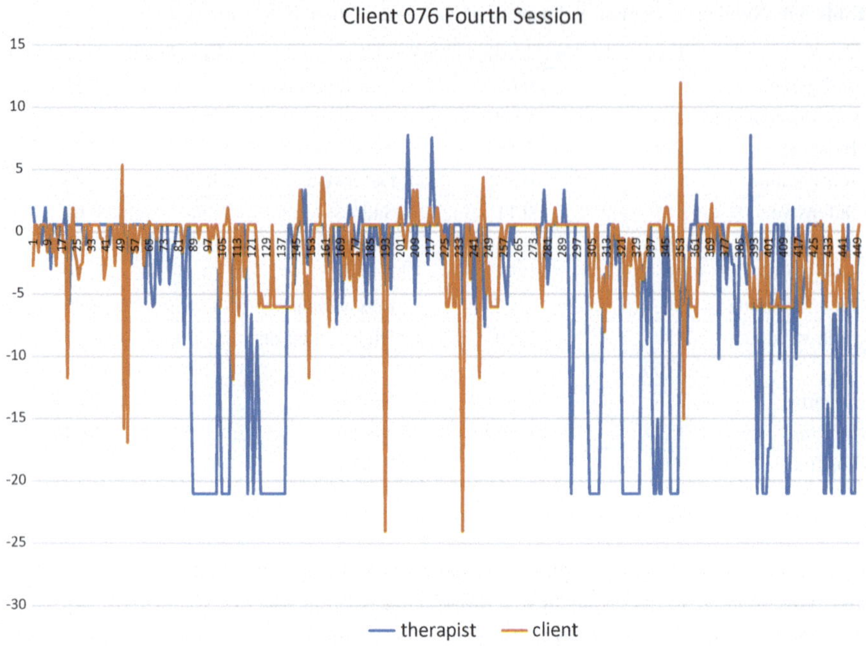

Fig. 3.6 Time series of the weighted SPAFF codes based on the fourth session of client number 076

belligerent may seem like an aggressive tactic, the therapist does not act belligerent to discourage the client. It is the intention of negative codes that is important here, not just the code itself. So when utilizing belligerence, the therapist is attempting to bring the client to understand that sometimes life's problems have no logical answer. The client at first turns away from this advice, but through the therapist's persistence, the client begins to understand the therapist's point of view. As mentioned previously, this case has had to move much more quickly than others, but the working alliance may have benefitted from this fast pace. The two have had to settle disagreements quickly to get to the "meat" of their sessions. Graphically (see Figs. 3.6 and 3.7), the therapist spends a lot of time with more displays of negative affect, but at the end of the encounter, the client winds up less negative than the therapist was, cumulatively (Fig. 3.7). The client, who had admitted no desire for friendship, brings up topics such as the meaning of life and spirituality when one would normally pose such questions toward like-minded friends. This shows the client has become fond of the therapist throughout their four sessions together. The client and therapist's match in personalities gave the client a sense of security to open up to the therapist as if they were a friend, further strengthening their alliance regardless of time constraints. However, what could not be ascertained by the video data was whether the course of therapy was successful in helping the client alleviate their symptom. Future research in this area will include measures of clinical outcomes.

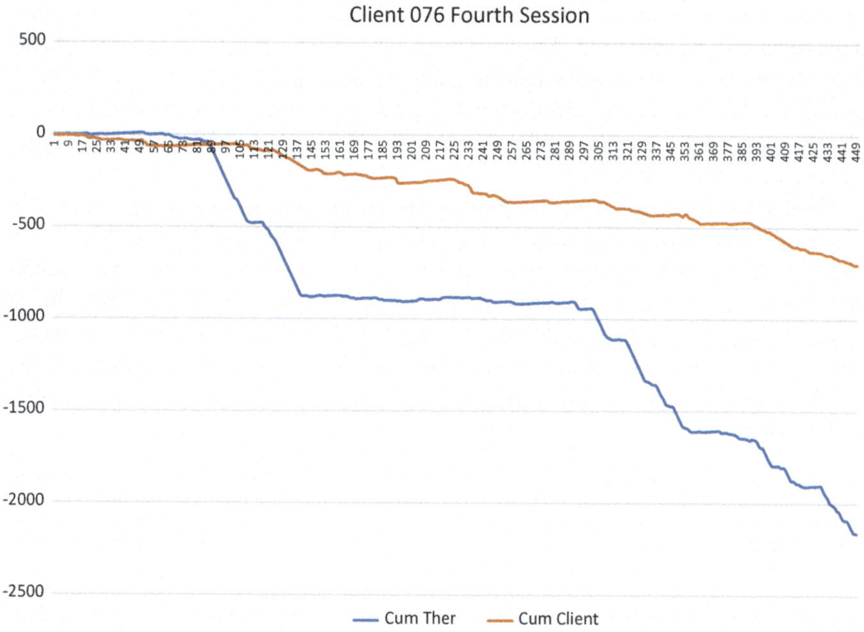

Fig. 3.7 Cumulative time series of the weighted SPAFF codes based on the fourth session of client number 076

3.5 Conclusion

Luedke (2013) found statistically significant differences in the relationships between the duration of therapist negative affect, the duration of therapist neutral affect, and client dropout. Clinicians who were more neutral ($M = 659$ vs. 470, $t(13) = -1.959$, $p < 0.05$) and less negative ($M = 56$ vs. 269, $t(13) = -2.065$, $p < 0.05$) in the last 15 min had better therapeutic outcomes. In the two cases presented here, using the *entire* session, the findings are similar: for the client who returned (042), the clinician was neutral over 75% of the time and negative less than 4% of the time, while the clinician for the client that did not return (031) was neutral 66% of the time and negative over 8% of the time. For clients, Luedke found a statistically significant relationship between the duration of negative affect coded for clients and premature dropout. Clients who were less negative returned to therapy ($M = 177$ vs. 313, $t(13) = -2.396$, $p < 0.05$). Again, this finding was echoed in the two clients presented here, as client 042 was negative 7.48% of the time, while client 031 was negative 25.37% of the time in the first session.

Taken together, these cases illustrate the utility and usefulness of SPAFF coding to describe the therapeutic relationship. Observational research provides a rich detailing of the affective exchange between the client and the therapist and provides a window on a process that is otherwise obscured and mysterious. At the same time, while the results of this investigation are both exciting and promising, several questions remain to be answered. The next logical step in this project is to target clients with the most common disorders, namely, anxiety and depression. In addition, research with client-therapist dyads that target anxiety and depression will allow for identification of those relationships that are related to a successful therapy and those that are not with these disorders.

Finally, Peluso et al. (2012) proposed using dynamical systems analysis to mathematically model the therapeutic relationship. These mathematical models can provide a rich graphical description of the dynamics of the relationship to clinicians and researchers alike. In addition, they provide testable hypotheses that clinical researchers can investigate. Specific patterns of affect dynamics can be correlated with treatment success measures such as retention, premature termination, and problem-specific outcome measures. This will be detailed in the following chapter in this volume.

References

Alexander, L. B., & Luborsky, L. (1987). The Penn helping alliance scales. In L. S. Greenberg & W. M. Pinsof (Eds.), *The psychotherapeutic process: A research handbook* (pp. 325–366). New York: Guilford Press.

Atzil-Slonim, D., Bar-Kalifa, E., Rafaeli, E., Lutz, W., Rubel, J., Schiefele, A., & Peri, T. (2015). Therapeutic bond judgments: Congruence and incongruence. *Journal of Consulting and Clinical Psychology, 83*, 773–784. https://doi.org/10.1037/ccp0000015

Burns, R. P., & Burns, R. (2008). *Business research methods and statistics using SPSS*. Sage Publications LTD. Retrieved from: http://www.uk.sagepub.com/burns/website%20material/Chapter%2025%20-20Discriminant%20Analysis.pdf

Crits-Christoph, P., Connolly Gibbons, M. B., Hamilton, J., Ring-Kurtz, S., & Gallop, R. (2011). The dependability of alliance assessments: The alliance-outcome correlation is larger than you might think. *Journal of Consulting and Clinical Psychology, 79*, 267–278. https://doi.org/10.1037/a0023668

Erzar, T., Rožič, T., Erzar, K. K., Čampa, B., & Čampa, V. M. (2012). Spremljanje čustvenih izrazov v terapevtskem procesu: priredba kodirnega sistema *SPAFF* (Specific Affect Coding System). / Adaptation of the Specific Affect Coding System (*SPAFF*). *Psihološka Obzorja/Horizons of Psychology, 21*(3–4), 47–56.

Falkenstrom, F., Granstrom, F., & Holmqvist, R. (2013). Therapeutic alliance predicts symptomatic improvement session by session. *Journal of Counseling Psychology, 60*, 317–328. https://doi.org/10.1037/a0032258

Fluckiger, C., Wampold, B. E., Del re, A., & Symonds, D. (2012). How central is the alliance in psychotherapy? A multilevel longitudinal meta-analysis. *Journal of Counseling Psychology, 59*, 10–17. https://doi.org/10.1037/a0025749

Gaston, L., & Marmar, C. R. (1994). The California psychotherapy alliance scales. *The Working Alliance: Theory, Research, and Practice*, 85–108.

Gaston, L., Marmar, C. R., Gallagher, D., & Thompson, L. W. (1991). Alliance prediction of outcome beyond in treatment symptomatic change as psychotherapy progresses. *Psychotherapy Research, 1*, 104–112. https://doi.org/10.1080/10503309112331335331

Gelso, C. J., & Carter, J. A. (1985). The relationship in counseling and psychotherapy: Components, consequences, and theoretical antecedents. *The Counseling Psychologist, 13*, 155–243. https://doi.org/10.1177/0011000085132001

Giese-Davis, J., Piemme, K. A., Dillon, C., & Twirbutt, S. (2005). Micro-variables in affective expression in women with breast cancer participating in support groups. In J. Harrigan, K. R. Scherer, & R. Rosenthal (Eds.), *Nonverbal behavior in the affective sciences: A handbook of research methods* (pp. 399–445). Oxford: Oxford University Press.

Gottman, J. M., McCoy, K., Coan, J., & Collier, H. (1996). The specific affect coding system. In J. Gottman (Ed.), *What predicts divorce: The measures* (pp. SPAFF 1–SPAF169). Hillsdale: Lawrence Erlbaum Associates, Inc.

Gottman, J. M., Woodin, E. M., & Coan, J. A. (1998). *Specific affect coding system manual (20-code version 4.0)*. Seattle: University of Washington.

Gottman, J. M., Murray, J. D., Swanson, C. C., Tyson, R., & Swanson, K. R. (2002). *The mathematics of marriage: Dynamic nonlinear models*. Cambridge, MA: The MIT Press.

Hersoug, A. G., Monsen, J. T., Havik, O. E., & Høglend, P. (2002). Quality of early working alliance in psychotherapy: Diagnoses, relationship and intrapsychic variables as predictors. *Psychotherapy and Psychosomatics, 71*, 18–27. https://doi.org/10.1159/000049340

Heyman, R. E. (2001). Observation of couple conflicts: Clinical assessment applications, stubborn truths, and shaky foundations. *Psychological Assessment, 13*(1), 5–35. https://doi.org/10.1037/1040-3590.13.1.5

Horvath, A. O., & Greenberg, L. S. (1986). The development of the working alliance inventory. In L. S. Greenberg & W. M. Pinsof (Eds.), *The psychotherapeutic process: A research handbook* (pp. 529–556). New York: The Guilford Press.

Horvath, A. O., Del re, A., Fluckiger, C., & Symonds, D. (2011). Alliance in individual psychotherapy. In J. C. Norcross (Ed.), *Alliance in individual psychotherapy* (2nd ed., pp. 25–69). New York: Oxford University Press.

Johnson, J. E., Burlingame, G. M., Olsen, J. A., Davies, D. R., & Gleave, R. L. (2005). Group climate, cohesion, alliance, and empathy in group psychotherapy: Multilevel structural equation models. *Journal of Counseling Psychology, 52*, 310–321. https://doi.org/10.1037/0022-0167.52.3.310

Lambert, M. J., & Barley, D. E. (2001). Research summary on the therapeutic relationship and psychotherapy outcome. *Psychotherapy: Theory, Research, Practice & Training, 38*(4), 357–361. https://doi.org/10.1037/0033-3204.38.4.357

Langhoff, C., Baer, T., Zubraegel, D., & Linden, M. (2008). Therapist-patient alliance, patient-therapist alliance, mutual therapeutic alliance, therapist-patient concordance, and outcome of CBT in GAD. *Journal of Cognitive Psychotherapy, 22*, 68–79. https://doi.org/10.1891/0889.8391.22.1.68

Luedke, A. (2013). Affect coding within the therapeutic relationship (Doctoral dissertation). Retrieved from http://purl.flvc.org/fcla/dt/3362484

Luedke, A. J., Peluso, P. R., Diaz, P., Freund, R., & Baker, A. (2017). Predicting dropout in counseling using affect coding of the therapeutic relationship: An empirical analysis. *Journal of Counseling & Development, 1995*, 125–135. https://doi.org/10.1002/j.1556-6676.2014.00000.x

Main, A. (2015). Empathic communication during mother-adolescent conflict management. Dissertation Abstracts International: Section B: The Sciences and Engineering, Vol 75(8-B(E)), No Pagination Specified.

Meier, P. S., & Donmall, M. C. (2009). Differences in client and therapist views of the working alliance in drug treatment. *Journal of Substance Use, 11*, 73–80. https://doi.org/10.1080/14659890500137004

Mozdzierz, G., Peluso, P. R., & Lisiecki, J. (2014). *Principles of counseling and psychotherapy: Learning the essential domains and nonlinear thinking of master practitioners* (2nd ed.). New York: Routledge.

National Institute of Mental Health. (2010). The numbers count: Mental disorders in America. Retrieved from: http://www.nimh.nig.gov/health/publications/the-numbers-count-mental-disorders-in-america/index.shtml

Norcross, J. C. (2010). The therapeutic relationship. In B. L. Duncan, S. D. Miller, B. E. Wampold, & M. A. Hubble (Eds.), *The heart and soul of change: Discovering what works in psychotherapy* (2nd ed., pp. 113–139). Washington D.C.: American Psychological Association.

Norcross, J. C. (Ed.). (2011). *Psychotherapy relationships that work: Evidence-based responsiveness* (2nd ed.). New York: Oxford University Press.

Norcross, J. C., & Lambert, M. J. (2011). Evidence-based therapy relationships. In J. C. Norcross (Ed.), *Psychotherapy relationships that work: Evidence-based responsiveness* (2nd ed., pp. 3–21). New York: Oxford University Press.

Norcross, J. C., & Wampold, B. E. (2011). Evidence-based therapy relationships: Research conclusions and clinical practices. *Psychotherapy, 48*(1), 98–102. https://doi.org/10.1037/a0022161

O'Malley, S. S., Suh, C. S., & Strupp, H. H. (1983). The Vanderbilt Psychotherapy Process Scale: A report on the scale development and a process-outcome study. *Journal of Consulting and Clinical Psychology, 51*, 581–586.

Peluso, P. R. (2015). *An Empirical Re-Weighting of the Specific Affect Coding System (SPAFF) for Therapist-Client Dyads.* Unpublished manuscript, Florida Atlantic University.

Peluso, P. R., Liebovitch, L. S., Gottman, J. M., Norman, M. D., & Su, J. (2012). A mathematical model of psychotherapy: An investigation using dynamic non-linear equations to model the therapeutic relationship. *Psychotherapy Research, 22*(1), 40–55. https://doi.org/10.1080/10503307.2011.622314

Sharf, J., Primavera, L. H., & Deiner, M. J. (2010). Dropout and therapeutic alliance: A meta-analysis of adult individual psychotherapy. *Psychotherapy Theory, Research, Practice, Training, 47*, 637–645. https://doi.org/10.1037/a0021175

Van Walsum, K. L. (2005). *Transference effects on student physicians' affective interactions and clinical inferences in interviews with standardized patients: An experimental study* (Doctoral dissertation). Retrieved from: http://repository.tamu.edu/bitstream/handle/1969.1/2548/etd-tamu-2005B-CPSY-van.pdf?sequence=1

Wampold, B. E. (2010). The research evidence for the common factors models: A historically situated perspective. In B. L. Duncan, S. D. Miller, B. E. Wampold, & M. A. Hubble (Eds.), *The heart and soul of change: Delivering what works in therapy* (2nd ed., pp. 49–81). Washington, DC: American Psychological Association.

Zandberg, L. J., Skriner, L. C., & Chu, B. C. (2015, April). Client-therapist alliance discrepancies and outcome in cognitive-behavioral therapy for youth anxiety. *Journal of Clinical Psychology, 71*, 313–322. https://doi.org/10.1002/jclp.22167

Zuroff, D. C., & Blatt, S. J. (2006). The therapeutic relationship in the brief treatment of depression: Contributions to clinical improvement and enhanced adaptive tendencies. *Journal of Consulting and Clinical Psychology, 74*, 130–140. https://doi.org/10.1037/0022-006X.74.1.130

Chapter 4
Dynamical Analysis of Therapist-Client Interactions

Paul R. Peluso, Andrew Z. Baker, Ashley Sauer, and Jennifer P. Peluso

4.1 Introduction

In a 2010 issue of *American Psychologist*, Vallacher, Coleman, Nowak, and Bui-Wrzosinska advocated for the use of dynamical systems for the modeling and study of complex human phenomena. The ultimate goal in creating and using such equations is to define the dynamics of a system and evaluate how the system changes over time (Gottman et al. 2002). This approach is common in *physical sciences* but rarely used in *social sciences* (Gottman and Notarius 2000, 2002), though this is beginning to change (see Vallacher et al. 2010). A dynamical systems approach uses (1) experience, data, and intuition as the starting point to develop a mathematical model, (2) mathematics to "solve" the model (i.e., determine its mathematical properties), and then (3) identified properties to make predictions which can be empirically tested. Insights resulting from applications of dynamical systems to social and psychological phenomena are not intended to represent the complexity of human interaction in its entirety, but they have the potential to reveal simple, yet crucial, dynamical features which underlie the emergence of complex behaviors in the system (see Nowak 2004).

Dynamic nonlinear mathematical equations allow investigators to deeply understand complex systems that are apt to change (Liebovitch et al. 2008). They can be used to determine the stable steady states or points of homeostasis within the system

P. R. Peluso (✉)
Department of Counselor Education, Florida Atlantic University,
Boca Raton, FL, USA
e-mail: ppeluso@fau.edu

A. Z. Baker · A. Sauer
The Alliance Lab, Department of Counselor Education, Florida Atlantic University,
Boca Raton, FL, USA

J. P. Peluso
Department of Psychology, Keiser University, Fort Lauderdale, FL, USA

U. Strawinska-Zanko, L. S. Liebovitch (eds.), *Mathematical Modeling of Social Relationships*, Computational Social Sciences,
https://doi.org/10.1007/978-3-319-76765-9_4

(i.e., the relationship). These steady states function as an anchor that brings the system back to homeostasis if the system is perturbed or if it has been moved away from homeostasis by a force. In regulated systems, these steady states can act as attractors. Gottman et al. (2002) utilized a metaphor of a rubber band to describe the properties of attractors, where the band, if stretched, snaps back into its original state when it is released. In essence, attractors define a "default" state for the system. Systems with stable steady states work in a similar fashion.

4.2 Mathematical Models: Using Dynamic Nonlinear Equations to Model Human Relationships

Over the last 15 years, dynamical systems approach has been used in social, developmental, and clinical psychology to describe human systems. In the latter context, Granic and her colleagues (Granic and Hollenstein 2003; Granic and Lamey 2002) have used dynamical systems to successfully model parent-child interactions and discover attractors in the system and other relational patterns that could guide intervention. Gardner et al. (2006) have suggested that dynamical systems may provide some traditional schools of family therapy with a mathematical tool to explain various theoretical components (e.g., homeostasis) as well as provide explanatory mechanisms for some quintessential family therapy techniques (e.g., paradoxical interventions, circular questioning, etc.). Dynamical systems have also proved highly useful in understanding conflicts between people, groups, and nations; such research has led to new translational methods to resolve intractable conflicts (Vallacher et al. 2013).

Gottman et al. (2002) realized the potential of investigating the dynamics of change beyond the marital dyad, stating:

> The model that was developed has given birth to a *theoretical language about the mechanism of change*. We did not have such a language before the model was successfully constructed. The model provides the language... suggests variables that can be targeted for change using interventions. In short, the model leads somewhere. It helps us raise questions, helps us wonder what the parameters may be related to and why. It raises questions of etiology...Thus, it is likely that the major contribution of the model will be the theoretical language and mathematical tools it provides. It will give us a way of thinking...that we never had before. (Gottman et al. 2002, p. 172)

Specifically, Gottman and his associates looked at several types of emotional behaviors displayed in a dialogue, including five positive codes (interest, validation, affection, humor, and joy), ten negative affect codes (disgust, contempt, belligerence, domineering, anger, fear/tension, defensiveness, whining, sadness, stonewalling), and a neutral affect code. Each of these includes specific verbal and nonverbal behavioral components for each emotional display (note: these were discussed in detail in Chap. 3).

Liebovitch and his associates (Liebovitch et al. 2008) modified Gottman's dynamic nonlinear equations to study the dynamics of conflicts between two parties (which could be individuals or groups). They determined how the dynamics of such conflicts and their emergent properties depended on the actions of each person. Taken together, these findings have guided our research. Our modeling of the therapeutic relationship constitutes a concrete extension of these results to a different social interaction. In this chapter, we will outline an ongoing research project (described in some detail in the previous chapter) that employs mathematical modeling to discover what underlies a successful therapeutic relationship. We will then discuss, through the use of case studies, the issues related to the parameters generated, the graphical displays of the derived models, and the indices of model fit.

4.3 Preliminary Analysis of Therapeutic Treatment: Mathematical Models

This section closely follows the work previously published by our group where we used a dynamical systems approach to mathematically model the therapist-client relationship. The rigorous mathematical derivation of the model was published by Liebovitch et al. (2011), and its application to psychotherapy was published by Peluso et al. (2012). Simulations were run that varied the strength of the client and therapist inertia parameter and the client and therapist influence parameter. This resulted in four basic models that can be used to predict positive and negative attractor states in the therapeutic relationship. The equations yielded scores that were the sum of two components: (1) interpersonal influence from therapist to client (and vice versa), termed "influenced behavior," and (2) each individual's (therapist's and client's) own dynamics, termed "uninfluenced behavior." The uninfluenced set point consisted of the emotional state (and baggage) that each person brings into the interaction, before being influenced by the other person. The influence component was defined as a function of one person's ability to move or change the other person's emotional state (positively or negatively) from the uninfluenced steady state. Ideally, the more positive steady states, and the fewer negative steady states that a relationship has, the more satisfying and regulated the relationship (Gottman et al. 2002).

4.3.1 The Equations of Therapy and Therapeutic Influence

Gottman et al. (2002) used difference equations to model affect as a function of time steps (intervals) between two people. Liebovitch and associates transformed those equations into differential equations to be able to use standard software for numerical integration and because there are powerful theorems and analytical methods available to analyze the dynamics of continuous systems that are not applicable to discrete systems (see, e.g., Liebovitch et al. 2008). The variables T(t) and C(t) denote the therapist's and client's behavior scores, respectively, at time t. The parameters a_T and a_C denote the inertia to change, that is, the degree of dependence of the therapist's and client's behavior scores on their previous values, respectively, where a_T, $a_C < 0$. (Formally, these parameters are the characteristic exponential relaxation times of the system rather than inertia related to a second derivative in time, but as they represent the influence of previous values on the current state, following the nomenclature used by Gottman et al. (2002), we refer to them as "inertia.") The smaller the absolute value of these parameters, the lesser the dependency on previous behavior scores. The parameters b_T and b_C denote the therapist's and client's behavior scores, respectively, when alone. The last part of the equations is the influence functions, which model the effect that the therapist or client has on the other person, that is, they are a function of either the client's or therapist's behavioral score. The functions $f_{C->T}(C)$ and $f_{T->C}(T)$ denote the form of the influence of the client on the therapist and the influence of the therapist on the client, respectively, and $s_{C->T}$ and $s_{T->C}$ denote the strength of those influences. The complete therapy equations are:

$$\frac{dT}{dt} = a_T T + b_T + s_{C \to} T f_{C \to} \mathrm{T}(\mathrm{C}) \tag{4.1}$$

$$\frac{dC}{dt} = a_C C + b_C + s_{T \to} C f_{T \to} C(T) \tag{4.2}$$

Translating these mathematical symbols into words, these equations state that the rate of change of the behavior state of the therapist and client (dT/dt and dC/dt) at time t is proportional to the sum of their inertia to change ($a_T T$ and $a_C C$), their behavior state when alone (b_T and b_C), and the influence from each other ($s_{C->T}$ $f_{C->T}(C)$ and $s_{T->C}$ $f_{T->C}(T)$). In other words, both the therapist and the client begin with an initial behavior score, and then these scores change over time based on the influence (or interaction) of the other person over time.

Based on clinical experience, and the research literature on psychotherapy, the research team created estimations of the influence function of the therapist and client (see Liebovitch et al. 2011, and Peluso et al. 2012, for a thorough discussion of the influence functions). Figure 4.1 displays an overlapping of the therapist's influence function and the client's influence function. This produces two steady states that are important to consider for the therapeutic relationship (see below for the more complete explanation in terms of nullclines). One is the steady state in the positive-positive space. It is a point where the therapist has more positive affect than the client. This seems to be the most beneficial spot for a therapeutic relationship to be in. At this point, the client is at least in the positive space but may either be trying to integrate new behaviors or new ways of thinking into his or her present life. Often, this instills a sense of hope, which is a precursor to change (Norcross 2011). If the therapist has been initially successful, then the therapist will have a good working relationship with the client to allow him or her to generalize the outcomes of their therapeutic work to other areas in the client's life. The second steady state is in the negative-negative space. Therapeutic relationships that move toward this attractor are in serious jeopardy of becoming entrenched in a fruitless struggle and eventually lead to a termination of the relationship. At this point, the therapist, despite previous efforts at using positive or neutral affect to elicit some positive

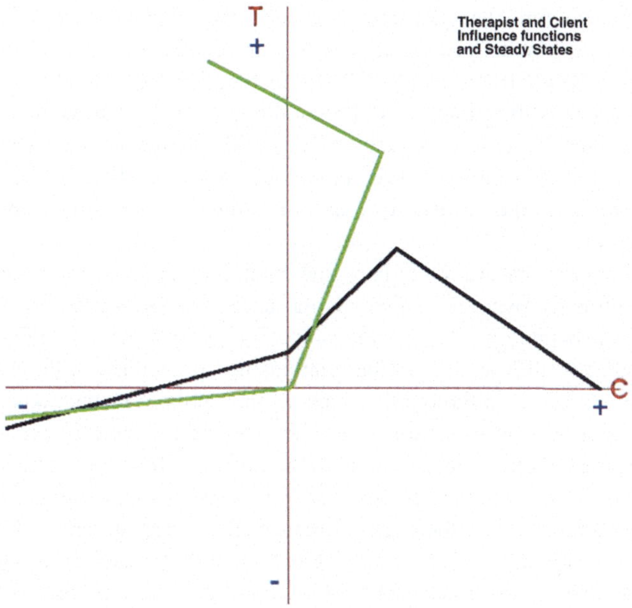

Fig. 4.1 Steady states in the therapeutic relationship

behavior change for a client, may—out of desperation or frustration—resort to negative emotion to provoke a reaction. However, this is often done without much reflection and is frequently ineffective.

4.3.2 Analysis of the Theoretical Model

The following steps present the analysis of the dynamics of the model: (1) first identifying the values of the behavior state of the therapist and client that define the critical points which can represent the final steady-state values that they reach at the conclusion of therapy, (2) then determining which values of the initial behavior state of the therapist and client will reach those final steady states, and (3) finally investigating how the dynamical behavior of the therapist and client depends on the parameters of the model.

4.3.3 Steady States in the Therapeutic Relationship

The behavior state of both the therapist (T) and client (C) starts from some initial values, and both evolve in time. We are very interested in determining where they wind up, as this represents the outcome of the therapy for both the therapist and client. Such steady states, if they exist, must occur where the values of T and C are no longer changing in time, namely, where the derivatives dT/dt = 0 and dC/dt = 0. The special values of T and C where this occurs are called the "critical points." The equations for dT/dt = 0 and dC/dt = 0 are each called the "nullclines." Since the critical points satisfy the equations of both nullclines, they lie at the intersections of the two nullclines. Therefore, we can easily see the values of these critical points graphically as the intersections of the nullclines on a plot where the behavior state of the therapist is on the vertical axis and the behavior state of the client is on the horizontal axis.

For the first case that we study here, the client and therapist each have an equal strength of influence over each other. As these are two "strangers" to one another, especially in the beginning of the relationship, there is no systematic influence of one over the other. The differentials in the relationship are handled in the weighting of the codes (see Chap. 3), although the shape of their influence functions is each different (as described above). Since, in this case, the "inertia" parameters are $a_T = a_C = -1$, the nullclines are the same as the influence functions. This means, that if we plot on the same graph (1) the function of how the therapist influences the client and (2) the function of how the client influences the therapist, where they intersect tells us the behavior states of the therapist and client that could represent their final steady states. This is illustrated in Fig. 4.1, where the two influence functions are plotted on a graph of the behavior state of the therapist versus the behavior state of the client. On that plot, we can see two intersections that identify two critical points.

Critical points can be stable or unstable. If small changes in T and C bring both the therapist and client back to their values at the critical point, then that critical point is a stable steady state. This is a point of equilibrium that the relationship would gravitate toward (or be attracted to). It is therefore called an attractor of the relationship. However, if small changes in T and C always push the therapist and client further away from the critical point, then it is unstable and does not represent a final steady state. We will see in the computer simulations below that the critical point at the intersection of the influence functions in the upper right of Fig. 4.1 is a stable steady state, while the critical point at the intersection of the influence functions in the lower left of Fig. 4.1.1 is unstable.

The stable critical point in the upper right of Fig. 4.1 is a good outcome, because both the therapist and client have positive behavior states for that stable steady state. The unstable critical point in the lower right of Fig. 4.1 is in a negative-negative space. We will see (next) that the dynamics near this critical point drag the client into even more negative behavior states. Therapeutic relationships that move in this direction are in serious jeopardy of becoming entrenched in a fruitless struggle, developing a therapeutic rupture, and potentially leading to a termination of the relationship.

4.3.4 Dynamics of the Therapeutic Relationship

We now continue the analysis of the model presented in Fig. 4.1 by determining the dynamics, that is, how the behavior state of the therapist and client evolves together in time. We do this by using a plot called a phase space. Each point in this phase space corresponds to one value T for the behavior state of the therapist on the vertical axis and one value C for the behavior state of the client on the horizontal axis. The therapist and client each start with an initial behavior state called their initial condition. The dynamics of the therapeutic relationship can then be visualized by following the path of this point, called its trajectory, as it moves through the plot, tracing out different values of the behavior state of the therapist and client.

We used computer software (MATLAB ODE113) to calculate how the values of T and C, for the behavior state of the therapist and client, respectively, in Eqs. (4.1) and (4.2), change in time when they start from many different initial conditions. We then plot these potential trajectories on the phase space. These simulations create a picture of how the therapeutic relationship might change given different initial conditions of the client and the therapist. These plots allow for a visual representation of the dynamics within the system and form the basis for predictions within the model (Liebovitch et al. 2011).

Figure 4.2 shows the phase-space plot of the trajectories of the therapist and client where the therapist and client have equal influence (meaning that the client and the therapist react emotionally to each other in an equal way) with each other. This corresponds to the same parameters and influence functions shown in Fig. 4.1. The dynamics shown in this phase space now reveals the nature of the two critical points identified by the intersection of the nullclines in Fig. 4.1. The lines (trajectories) in Fig. 4.2 clearly show that many initial conditions are attracted to a stable endpoint at the critical point in the upper right corner. On the other hand, for the trajectories avoid-

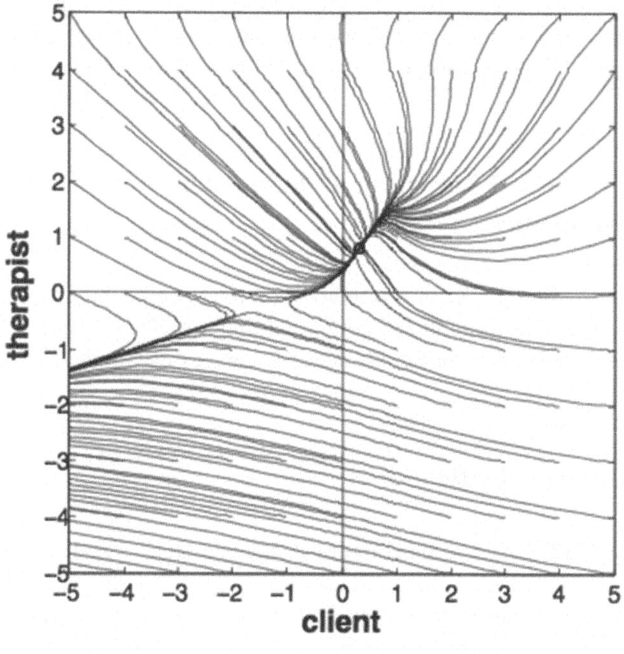

Fig. 4.2 Phase state plot

ing the critical point in the lower left of the point, it is an unstable critical point (a saddle). It is important to note that it is the final steady states reached by the client—determined by the long time solution of the differential equations—that the model uses to evaluate the success of therapy, rather than the transient dynamics. We can use an ordinary differential equation model to investigate the role of transient dynamics (e.g., transient negative affects from the therapist being used to challenge the client).

In this instance, where client and therapist have equal influence with each other, the relationship will likely end up in at this positive attractor, as long as the therapist begins with positive emotion. However, if the client starts therapy in a very negative emotional state (−5 or −4), then the therapist must be more positive in order to overcome the movement toward the negative state. Furthermore, in this model, if the client begins therapy with a mild negative state (−1) or is neutral, the therapist can also match the negative emotion and still attract the relationship toward the positive stable steady state (approximately +1 for therapist and client). In addition, if the client starts therapy with very positive affect (+2, to +5), the therapist can also display some negative or neutral emotion and still draw the relationship to the positive steady state. "Going negative" can be a strategy for the therapist to either bring a client who is mildly negative or neutral (−1 or 0) about therapy into a positive space. It may also be a strategic method for "tamping down" a client who is displaying highly (and possibly unrealistically) positive emotions. Since, in this scenario both the client and the therapist are equally influential of the other, if the therapist initially is highly positive (+5 to +2) and the client is either negative or positive, the therapist will be drawn down toward the positive stable steady state. The key, it seems, for this relationship is that (in most instances) the therapist must avoid beginning a negative affect. The only exception is if the client is initially very negative, in which case, the relationship will be pulled toward negative emotional states (Peluso et al. 2012).

4.4 Mathematical Modeling of Actual Therapist-Client Relationships

The previous chapter presented information on the Specific Affect Coding System (SPAFF) in detail. For more information regarding the SPAFF codes, the reader is directed to review the material (Peluso et al. Chap. 3, this volume).

4.4.1 Parameter Estimation

Finally, the SPAFF data was used to create mathematical models of the recorded first session of the therapist-client relationships. In this limited data set, the dependent variable was whether the *client dropped out* after one session or if the *client returned* four or more times (suggesting that a successful therapeutic relationship was established). Initially, we ran three (3) separate mathematical models using the "dyad" program created for the Gottman Lab (available in the CRAN Repository, cran.r-project.org). This program, written in the R programming language, can run a model that fits a "bilinear" influence function to the data. This means that the influence function is two separate line segments that are joined at a threshold point. It creates parameters for both therapist and client that include the slopes of the two segments of the influence function, the threshold where the line segments change, the initial state of the model, and the inertia within the model. A set of parameters is generated for both the client and the therapist.

The second model that is run is the "combination" model, which combines the bilinear model described above with an "ojive" model suggested by Gottman et al. (2002). In the combination model, the influence function is created in the same way as the bilinear model, but the line segments are not necessarily contiguous at the threshold point. This model allows for the influence function to account for larger changes between the two segment areas and (theoretically) created a better fit for the data. The same parameters are generated, with an additional parameter to account for the change in the two line segments for the influence function. Finally, there is a "no-influence" model, which is an autoregressive model where there is no influence modeled. As a result, only the initial state and inertia parameters are created.

4.4.2 Model Validation

A frequent problem that researchers using dynamical systems have faced is estimating the parameters from the data and then applying the best fit to the data. Models can be created, but when there is no way to evaluate their goodness of fit to real data, then the models are illustrative and speculative at best but may not hold up to the most basic principle of science: falsifiability. Recently, Hamaker and her colleagues (2009) devised a method for estimating the parameters for a model and computed both a Bayesian information criterion (BIC) and an Akaike information criterion (AIC). These can be used to compare models using different influence functions and to determine the best fit for the given interaction. Each method for assessing the goodness of fit in a model places different penalties for the number of parameters in the model, as well as different underlying assumptions about whether a "true" model exists (see Hamaker et al. 2009, for a discussion). Since the combination model has the most parameters, and the no-influence model has the least, it will be important to consider both of these fit indices to insure that a well-fitting model is not rejected (Type 2 error). In fact, Madhyastha et al. (2011) have recently used both models to test Gottman et al.'s (2002) original influence functions. While they included the "no-influence" model for comparison purposes, we have chosen not to incorporate that for analyses. This was driven by two theoretical reasons. First, the no-influence model has only three (3) parameters, while the Bilinear model has six (6), and the Combination model has seven (7). Therefore, the no-influence model will always be superior statistically but perhaps not the best theoretically. The

Table 4.1 Model fit statistics comparison

	BIC				AIC			
	Therapist		Client		Therapist		Client	
Case	Bilinear	Combi	Bilinear	Combi	Bilinear	Combi	Bilinear	Combi
005A	1599.46	1575.43	1798.09	1792.19	1556.26	1549.52	1766.89	1766.12
042A	2520.14	2522.73	1716.59	1722.52	2495.49	2493.98	1691.96	1693.77
042B	2784.29	2789.10	1835.98	1834.40	2759.65	2760.36	1811.34	1805.64

Note: "A" denotes first session, "B" denotes fourth session

second (related) theoretical reason is that there is an influence between therapists and clients, and it does not make sense to model the data without including an influence function. These goodness of fit tests are also a part of the "dyad" program and are computed at the same time as the parameters for the models are estimated. Thus, employing this approach of parameter estimation and information criteria will help to determine the best fitting model for therapists and clients, which may solve the problem of determining the effect of therapist responsiveness (differing levels of affective involvement that emerges within the context of client needs) on client outcomes.

A review of Table 4.1 shows the fit statistics for three model runs (005A, 0042A, and 0042B, which will be described below). Specifically, the BIC for the therapists and clients comparing the bilinear model to the combination model and the AIC for therapists and clients comparing the bilinear model to the combination model are included. According to Kass and Wasserman (1995), models with the lower BIC or AIC are considered to be better fits than models with higher numbers. However, there are two other factors to consider with model fit: the underlying assumptions of the fit indices and the magnitude of the differences in the models.

Both BIC and AIC are based on information theory, which estimates the amount of "information" that is lost when a model is fit to the data that created it (Burham and Anderson 2004). This is used to select the best from among multiple models (and parameters) for retention and further use. The difference between BIC and AIC is that the basic assumption of the BIC model fit statistics is that there is one "true" model, while AIC is designed to find the best fitting model (Acquah 2010). In addition, the BIC generally penalizes free parameters more strongly than the AIC, thus emphasizing simpler models, rather than more complex ones. Burham and Anderson (2004) recommend AIC for models that use real data collected from a naturalistic sample, like the kind of observational approach used here.

According to Kass and Rafferty (1995), any differences less than three are "not worth mentioning," while differences of ten (10) or more are considered very strong evidence that one model is better than the other (Burnham and Anderson 2004). An investigation presented in Table 4.1 shows that for the therapists, the differences in BIC between the bilinear model and combination model are 24 (in favor of the combination model), 2, and 5 (in favor of the bilinear model, and for clients are 6 in favor of the combination model), 6 (in favor of the bilinear model), and 1. So while the bilinear model BIC values are lower in two out of the three cases, they are not in the very strong evidence category. Looking at the AIC values for the therapists, the combination model is lower for two of the three cases (both by less than 2), and bilinear model is lower by 2 as well. Differences in the models for clients were all less than 2 in two out of the three and less than 6 in the other. Therefore, given the small magnitude of difference, particularly with AIC, concerns of "overfitting" that come with adding the additional parameter in the combination model, compared to bilinear model, do not appear. In addition, given that the AIC is preferred with real-world data, and looks for the best fitting model (versus a "true" model), it would seem that the combination model is preferable and, on further analysis, that the AIC is an effective approach for ascertaining the best fitting model. As a result, we will continue with the combination model for our presentation below.

We now present three therapy sessions from two separate case studies to illustrate the mathematical modeling, the difference in the model parameters, and a graphical representation of each.

4.4.3 Unsuccessful Therapeutic Relationship

The first client that we present was an example of unsuccessful therapy. Figure 4.3 shows the influence function for the combination model of a client that did not return to therapy after the first session, while Fig. 4.4 shows the phase portrait for the initial therapy session of the same client for the combination model. In Fig. 4.3, the client's influence on the therapist is the red line, while the therapist's influence on the client is the blue line. The axes in the figure show the relative emotional valence (from negative to positive) for the therapist (ordinate) and client (abscissa) and are derived from the weightings of the SPAFF affect codes. As with Fig. 4.1, the areas of intersection of these influence functions indicate where a steady state or attractor point will be for the system. In this instance, there is one main intersection in the lower left-hand quadrant (therapist negative, client negative). What is noteworthy here is that the therapist's influence on the client is relatively flat and negative, suggesting that the therapist was not very influential overall; whatever influence did exist tended to be negative.

Figure 4.4 is the phase-space portrait created in the MATLAB program, which incorporated the model parameters derived by the "dyad" program. Each of the two axes is the relative emotional valence based on the weightings of the SPAFF affect codes. As with Fig. 4.3, the steady-state attractors can be seen, as well as the relative influence of the attractors on the entire system. In addition, the phase-space portrait helps to tell the story of the relationship between the therapist and client by tracing the trajectory of the relationship, depending on the starting point. In this instance, there are two areas of attraction at coordinates (−6, −7) and (−5, −6)—and there is an attractor point (or saddle node) around coordinates (3, −7). Tables 4.2 and 4.3 portrays the coordinates of the first attractor point at the intersection of the threshold where the first regime (and line of fit) changes and the second regime begins. So in this situation, even if the client started out slightly positive (+5) and the therapist

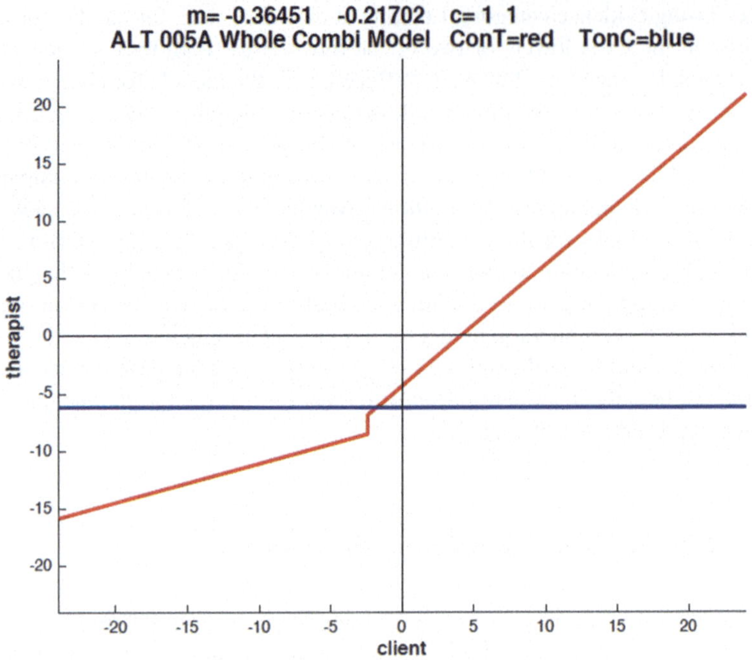

Fig. 4.3 Influence functions (negative outcome)

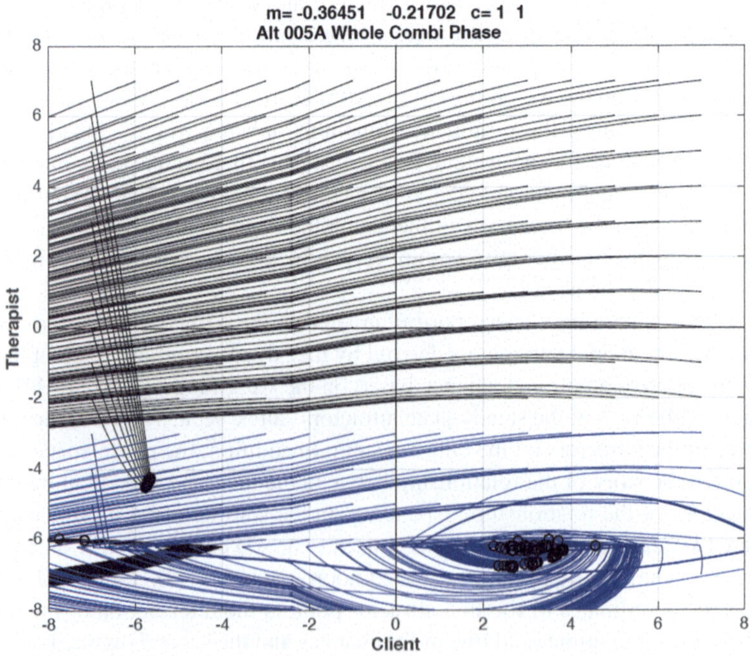

Fig. 4.4 Phase-space portrait (negative outcome)

Table 4.2 Combination model parameters for therapists

Case	Initial state	Inertia	Difference in initial state	Left slope	Right slope	Threshold
005A	−2.5142	0.63549	1.8621	0.12286	0.38398	−2.4
042A	4.3723	0.318684	−3.92968	−0.01288	−0.90075	2
042B	0.73413	0.29984	1.562404	0.587669	0.012713	−0.2

Table 4.3 Combination model parameters for clients

Case	Initial state	Inertia	Difference in initial state	Left slope	Right slope	Threshold
005A	−0.9966	0.78298	−8.7695	−0.87711	0.02113	−6.2
042A	1.6014	0.16631	−1.08379	−0.407151	−0.063163	5.4
042B	3.6875	0.23755	−3.1042	0.00037	−0.21425	8.2

started out more positive (+10), the relationship would still be attracted toward the steady state in the lower left-hand (or "negative-negative") quadrant. There seems to be a slim chance of the relationship being attracted to the saddle node, and the "best-case" scenario for this therapeutic dyad is the attractor point in the lower right-hand quadrant. None of these results are very positive for the therapeutic relationship, and (as mentioned above) the client did not return to therapy, despite the fact that the client had made a follow-up appointment with the therapist—a common problem in therapy.

4.4.4 Successful Therapeutic Relationship

We present a contrasting mathematical model for a successful therapeutic encounter (*note:* this is the mathematical modeling of a client whose time series charts were displayed in Chap. 3). Figure 4.5 presents the influence functions derived from a different client's first therapy session. There is one area of intersection between the client's influence function and the therapist's influence function in the lower right-hand quadrant (the "negative/positive" space). Figure 4.6 presents the phase-space portrait of the model. Four critical points are evident: two in the positive/positive space and two in the negative/positive space. Like the unsuccessful client above, some of these critical points may be saddle nodes where either $dT/dt = 0$ or $dC/dt = 0$. Nonetheless, their close proximity shows two regions to which the relationship will be attracted, with the client being more positive than the therapist and/or where the client is positive and the therapist is slightly negative. As with Figs. 4.2 and 4.4, we can trace the trajectories from a hypothetical starting point and see that the relationship (based on the dynamics derived from the best fitting mathematical function) will be attracted toward one of the two positive attractor points in the upper right-hand quadrant. For example, if the client started out slightly negative (−2) and the therapist started out more negative (−4), then the relationship would be attracted toward the steady state in the upper left-hand (or therapist positive/client

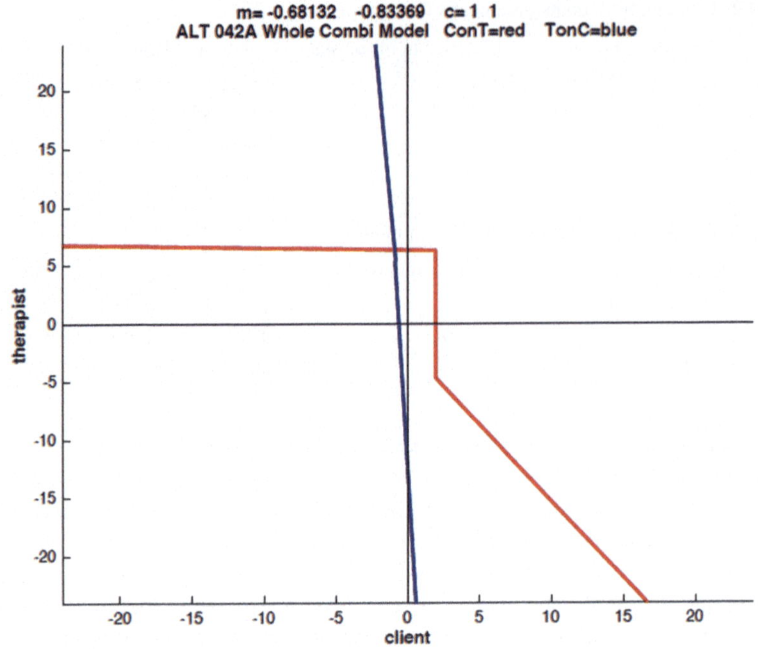

Fig. 4.5 Influence function (positive outcome, first session)

negative) quadrant. However, if the therapist started more positive (+6) and the client remained negative (−4), the relationship would also be attracted toward the other steady state in the upper left-hand quadrant (therapist positive/client negative), but with the therapist more positive. This outcome is not clearly positive but is often seen in successful therapeutic interactions, based on previous data (see Peluso et al. 2012). The client persists in therapy and continues productive work.

This case, however, illustrated a therapeutic relationship that was successful (i.e., the client continued in therapy past the fourth session), and we can look at how the relationship changes over the course of time from the first and the fourth session. Figure 4.7 shows the influence functions for the therapist and client. What is noteworthy is that the intersection of the influence functions has moved from the upper right-hand quadrant to the lower left-hand quadrant (client positive, therapist negative). This suggests that, at this point in the therapeutic endeavor, the therapist's influence is positive, but dynamically so. In practical terms, the therapist has a constant and positive influence on the client, but not one that is affected by the client's positive or negative affect. This is directly opposite to the case presented previously, where the therapist's influence was negative but flat (suggesting that there was little direct influence). In Fig. 4.8, the phase portrait has an attractor in the lower right-hand quadrant (client positive/therapist negative), as expected.

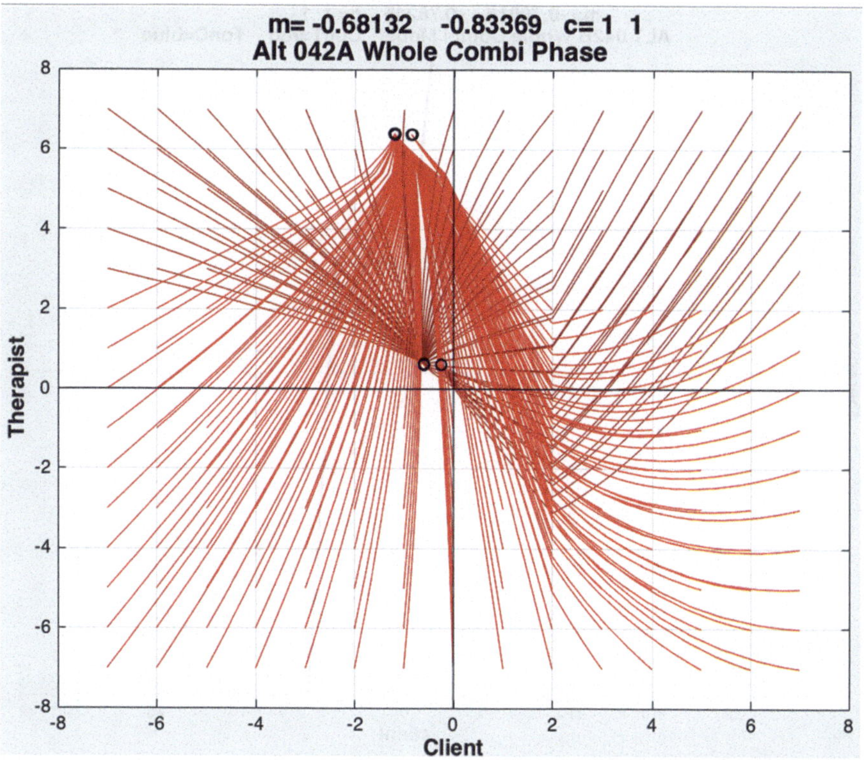

Fig. 4.6 Phase-space portrait (positive outcome, first session)

In this situation, the mathematical models, like the one shown in Figs. 4.5 and 4.6, were indicative of the models that were found with returning clients having steady states in the lower right-hand quadrant. Future studies will need to correlate these models (and their parameters and outcome measures) to see if, in addition to returning or dropping out, the placement of the attractors is correlated with clients who report significant symptom reduction.

Another interesting overall finding from our preliminary analyses revealed that several individual SPAFF codes were important in differentiating successful and unsuccessful therapeutic relationships. Specifically, domineering and tension were indicative of less productive relationships, while high validation and interest were found in more productive relationships. In addition, relationships with a higher percentage of neutral codes were correlated with a successful relationship. These findings may be the beginning of treatment recommendations that can be fed back to clinicians.

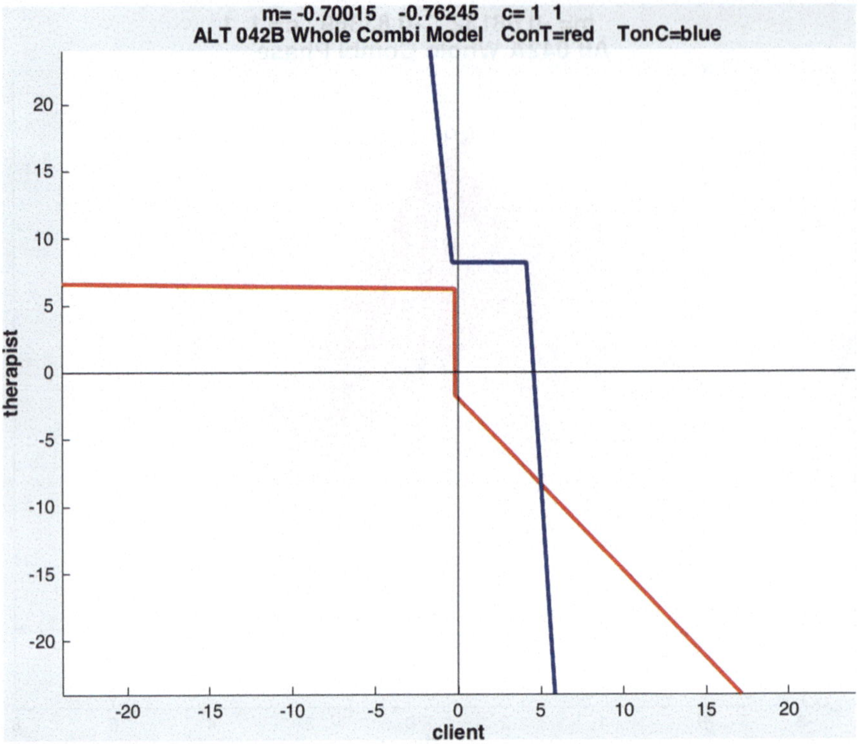

Fig. 4.7 Influence function (positive outcome, fourth session)

4.5 Future Directions

While the results of this investigation are both exciting and promising, there are several questions that remain to be answered. The next logical steps in this project are to record sessions that span an entire course of treatment for mathematical modeling and also to apply our current findings to treatment for specific disorders, for example, anxiety and depression. In addition, it will be important to correlate observed parameters and model features with clinical outcome measures (e.g., symptom reduction, goal attainment, etc.). Lastly, following the research lines that Gottman and his colleagues pioneered with husbands and wives, we wish to add physiological measures of arousal to our methods to determine their separate and additive impact on the therapeutic relationship.

Several interesting sets of post hoc analyses will also be possible as a result of demographic variations among our therapists and clients (e.g., race/ethnicity and sex). Another subgrouping will concentrate on the skill level of the therapists observed (based on number of years of practice). By analyzing the recordings of sessions of both novices and experts, researchers can compare how each group of

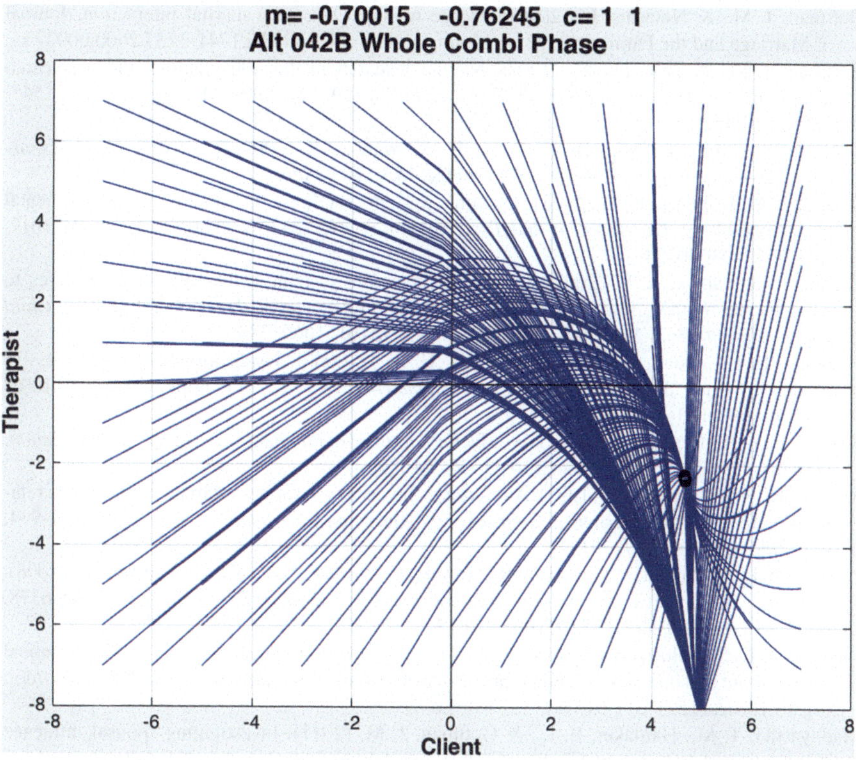

Fig. 4.8 Phase-space portrait (positive outcome, fourth session)

counselors create and maintain the therapeutic relationship with clients at several different starting points and time points. Measures of client and therapist satisfaction with the session and relative strength of the therapeutic alliance will also be collected for comparison. However, since these will not be demographic factors specifically controlled for, the analyses will be exploratory in nature.

References

Acquah, H. G. (2010). Comparison of Akaike information criterion (AIC) and Bayesian information criterion (BIC) in selection of an asymmetric price relationship. *Journal of Development and Agricultural Economics, 2*(1), 1–6.

Burnham, K. P., & Anderson, D. R. (2004). Multimodel inference: Understanding AIC and BIC in model selection. *Sociological Methods & Research, 33*, 261–304. https://doi.org/10.1177/0049124104268644

Gardner, B. C., Burr, B. K., & Wiedower, S. E. (2006). Reconceptualizing strategic family therapy: Insights from a dynamic systems perspective. *Contemporary Family Therapy, 28*, 339–352. https://doi.org/10.1007/s10591-006-9007-x

Gottman, J. M., & Notarius, C. (2000). Decade review: Observing marital interaction. Journal of Marriage and the Family, 62, 927–947. https://doi.org/10.1111/j.1741-3737.2000.00927.x

Gottman, J. M., & Notarius, C. (2002). Marital research in the 20th century and a research agenda for the 21st century. *Family Process, 41*(2), 159–197. https://doi.org/10.1111/j.1545-5300.2002.41203.x

Gottman, J., Murray, J., Swanson, C., Tyson, R., & Swanson, K. (2002). *The mathematics of marriage: Dynamic nonlinear models*. Cambridge, MA: MIT Press.

Granic, I., & Hollenstein, T. (2003). Dynamic systems methods for models of developmental psychopathology. *Development and Psychopathology, 15,* 641–669. https://doi.org/10.1017. S0954579403000324

Granic, I., & Lamey, A. V. (2002). Combining dynamic systems and multivariate analyses to compare the mother-child interactions of externalizing subtypes. *Journal of Abnormal Child Psychology, 30*, 265–283. https://doi.org/10.1023/A:1015106913866

Hamaker, E. L., Zhang, Z., & Van der Maas, H. L. J. (2009). Using threshold autoregressive models to study dyadic interactions. *Psychometrika, 74*, 727–745. https://doi.org/10.1007/s11336-009-9113-4

Kass, R. E., & Raftery, A. E. (1995). Bayes factors. *Journal of the American Statistical Association, 90*(430), 773–795. https://doi.org/10.2307/2291091

Kass, R. E., & Wasserman, L. (1995). A reference Bayesian test for nested hypotheses and its relationship to the Schwarz criterion. *Journal of the American Statistical Association, 90*, 928–934. https://doi.org/10.2307/2291327

Liebovitch, L. S., Naudot, V., Vallacher, R., Nowak, A., Bui-Wrzosinska, L., & Coleman, P. (2008). Dynamics of two-actor cooperation-competition conflict models. *Physica A, 387*, 6360–6378. https://doi.org/10.1080/10781911003691625

Liebovitch, L. S., Peluso, P. R., Norman, M. D., Su, J., & Gottman, J. M. (2011). Mathematical model of the dynamics of psychotherapy. *Cognitive Neurodynamics, 5*(3), 265–275. https://doi.org/10.1007/s11571-011-9157-x

Madhyastha, T. M., Hamaker, E. L., & Gottman, J. M. (2011). Investigating spousal influence using moment to moment affect data from marital conflict. *Journal of Family Psychology, 25*(2), 292–300. https://doi.org/10.1037/a0023028

Norcross, J. C. (2011). *Psychotherapy realtionships that work: Evidence based responsiveness* (2nd ed.pp. 3–21). New York: Oxford University Press.

Nowak, A. (2004). Dynamical minimalism: Why less is more in psychology. *Personality and Social Psychology Review, 8*(2), 183–192. https://doi.org/10.1207/s15327957pspr0802_12

Peluso, P. R., Liebovitch, L. S., Gottman, J. M., Norman, M. D., & Su, J. (2012). A mathematical model of psychotherapy: An investigation using dynamic non-linear equations to model the therapeutic relationship. *Psychotherapy Research, 22*(1), 40–55. https://doi.org/10.1080/10503307.2011.622314

Vallacher, R. R., Coleman, P. T., Nowak, A., & Bui-Wrzosinska, L. (2010). Rethinking intractable conflict: The perspective of dynamical systems. *American Psychologist, 65*(4), 262–278. https://doi.org/10.1037/a0019290

Vallacher, R. R., Coleman, P. T., Nowak, A., Bui-Wrzosinska, L., Liebovitch, L., Kugler, K., & Bartoli, A. (2013). *Attracted to conflict: Dynamic foundations of destructive social relations dynamic foundations of destructive social relations*. Berlin, Heidelberg: Springer.

Chapter 5
Modeling Psychotherapy Encounters: Rupture and Repair

Urszula Strawinska-Zanko, Larry S. Liebovitch, and Paul R. Peluso

5.1 The Therapeutic Relationship

According to the National Institute of Mental Health, one out of every four adults in the United States suffers from a diagnosable mental illness (NIMH 2010). Researchers have found that psychotherapy is an effective approach for treating mental illnesses and that the therapeutic relationship is a key component in the success of therapy (Duncan et al. 2010; Mozdzierz et al. 2014; Norcross 2010; Wampold 2010). Norcross and Wampold (2011) summarized the findings of the American Psychological Association's task force on evidence-based therapy relationships, which affirmed Lambert's finding (Lambert 1992; Lambert and Barley 2002) that the therapeutic relationship accounts for as much of the variance accounted for in successful therapy as the type of method or approach. In addition, they suggested a need to focus on specific therapist behaviors that would promote a positive therapeutic relationship.

U. Strawinska-Zanko (✉)
Department of Conflict Resolution Studies, College of Arts, Humanities, and Social Sciences, Nova Southeastern University, Fort Lauderdale, FL, USA
e-mail: uzanko@nova.edu

L. S. Liebovitch
Departments of Physics and Psychology, Queens College, City University of New York, Flushing, NY, USA

Physics Program, The Graduate Center, City University of New York, New York, NY, USA

Adjunct Senior Research Scholar, Advanced Consortium for Cooperation, Conflict, and Complexity (AC4), The Earth Institute at Columbia University, New York, NY, USA

P. R. Peluso
Department of Counselor Education, Florida Atlantic University, Boca Raton, FL, USA

U. Strawinska-Zanko, L. S. Liebovitch (eds.), *Mathematical Modeling of Social Relationships*, Computational Social Sciences,
https://doi.org/10.1007/978-3-319-76765-9_5

Liebovitch et al. (2011) and Peluso et al. (2012) set out to utilize a dynamical systems approach, to look at the complex relationship attributes between therapist and client. Building on the work of Gottman and his associates (Gottman et al. 2002), and Liebovitch et al. (2008), they developed a mathematical model of the therapeutic relationship based on clinical and research findings. In this chapter, we wish to present an extension to that work by focusing on an additional element of the therapeutic relationship that can negatively impact the success of therapy: therapeutic ruptures and the attempts to model a repair to them.

5.2 Therapeutic Ruptures and Repairing Therapeutic Ruptures

As important as the therapeutic relationship is to the successful outcome of therapy, sometimes, breaks or ruptures to the therapeutic alliance can occur. According to Safran et al. (2002), a rupture in the therapeutic alliance is defined as "a tension or breakdown in the collaborative relationship between patient and therapist" (p. 236). Such breaks can be precipitated by any number of factors, such as failing to engage and connect with the client, assess their needs, or agree on goals (Mozdzierz et al. 2014). There are three major sources of ruptures: differences between therapist and client about the tasks of therapy, differences between them about the goals of treatment, and "strains" in their connection (Safran et al. 2002). Bordin (1994) suggested that within the therapeutic relationship, "negotiations" were constantly being conducted between client and therapist. These negotiations concern the three major areas (i.e., tasks, goals, and bond) in two dimensions: a conscious level (i.e., surface meaning) and unconscious level (i.e., underlying meaning). Disagreements may be expressed explicitly or implicitly and have the potential to sabotage the therapeutic endeavor altogether (Mozdzierz et al. 2014).

Some research findings related to therapeutic ruptures suggest that clients frequently have negative feelings about the therapist or therapy but do not give voice to them (Coutinho et al. 2011). Often, ruptures are a reoccurrence of previous ruptures that have not been dealt with, or may result from a situation when a client was unprepared to respond to a therapeutic suggestion, or when feelings of confusion or ambivalence were present. In fact, it is generally not until the end of the therapeutic endeavor, when a client terminates therapy, that most will express any negative emotions toward the therapist (Coutinho et al. 2011).

The two types of behaviors that seem to indicate that a rupture has occurred are withdrawal behaviors and confrontation behaviors (Safran and Muran 2000). Withdrawal behaviors are any actions that seem to limit the client's participation in therapy and signal his or her disengagement. Confrontational behaviors are expressions of anger, frustration, or resentment toward the therapist or the therapeutic process. Each of these types of behaviors is designed to stop the therapeutic process and force the therapist to react. The therapist's reaction will generally determine whether the rupture will become a breach in the alliance (with the potential to effectively end the therapy) or will be repaired in an attempt to salvage (and possibly strengthen) the alliance (Mozdzierz et al. 2014). Safran and his colleagues (2002) developed a conceptually useful four-stage framework for repairing alliance ruptures that is effective in maintaining the therapeutic relationship. Briefly, these stages are (a) attending to the rupture behavior (i.e., withdrawal or confrontation), (b) exploring the rupture experience, (c) exploring the client's avoidance, and (d) the emergence of a wish or need.

5.3 Gottman's Relationship Repairs

Gottman and his associates (Gottman et al. 2002) in their work with marital relationships found a similar dynamic to Safran et al.'s (2002) rupture repairs in successful relationships:

> The most consistent discriminator between distressed and non-distressed marriages that has been obtained across four nations is negative affect reciprocity. Negative affect reciprocity is the sequential pattern that negative affect by a spouse is more likely than his or her baseline after the partner has just been negative. In other words, for unhappy couples, negative behavior by the wife or husband generally produces a negative response in the spouse. This implies that it is very important that a couple be able to repair the interaction and exit a negative affect cycle, which may be the sine qua non of effective marital interaction during conflict. This view is further supported by the fact that happy couples do exhibit negativity—even a significant quantity. The predictors of divorce...are not unknown in marriages that are stable and happy. The crucial difference appears to lie in the couples' ability to repair their negativity, not the ability to avoid it entirely. (Gottman et al. 2002, pp. 189–190)

According to Gottman et al. (2002), a repair attempt should have two characteristics: it should be asymmetrical, that is, it should be activated once negativity has exceeded a certain threshold, and it should appear abrupt once the threshold has been crossed. Gottman (2011) found through performing a hidden Markov analysis that in the case of unhappy couples, negative affect is an "absorbing state" that is easy to enter from neutral or positive affect and hard to exit. According to Gottman, conflicts become similar to stepping into a quicksand bog: no matter how one perturbs the system, one sinks deeper into more negativity. He describes it as the "Roach Motel," where couples "check into negativity but they don't check out." These conditions appear to be similar to what therapists encounter with a therapeutic rupture (especially one that is not resolved appropriately). These characteristics will be applied to Liebovitch et al. (2011) and Peluso et al.'s (2012) model of the therapeutic relationship to determine the theoretical changes in a dynamical systems model for the therapeutic relationship.

5.4 The Mathematical Model of the Therapeutic Relationship

In this chapter we are presenting an extension to the mathematical model of psychotherapy developed by Liebovitch and Peluso (Liebovitch et al. 2011; Peluso et al. 2012) by introducing a new component that models the repair term. We are interested in investigating what effect the repair attempt has on the outcome of psychotherapy. We expect that there might be circumstances when a display of negative emotion may be beneficial to the therapeutic relationship. For example, appropriate confrontation or expressions of disappointment may be a necessary feedback to the client (Coutinho et al. 2011). Again, the immediate result may be a therapeutic rupture, but if it is done purposefully or strategically, it may have a long-term benefit for the client. In the next section, we explore how repair attempt and its strength and timing affect the therapeutic relationship.

In the following paragraphs, the original model (Liebovitch et al. 2011; Peluso et al. 2012), as well as the extended model, will be characterized. In the subsequent sections, the simulation work with the repair term will be presented and discussed. Possible practical implications and further directions conclude the chapter.

Liebovitch and Peluso (Liebovitch et al. 2011; Peluso et al. 2012) modeled the psychotherapeutic relationship as a set of nonlinear two-dimensional ordinary differential equations (ODEs) (see Chap. 4 in this volume for the model description and its variants).

$$\frac{dT}{dt} = m_1 T + b_1 + c_1 F_C(C) \quad F_C(C) = \begin{cases} 0.5C + 0.5 & C \le 0 \\ C + 0.5 & 0 < C \le 1 \\ -0.5C + 2 & C > 1 \end{cases}$$
$$\frac{dC}{dt} = m_2 C + b_2 + c_2 F_T(T) \quad F_T(T) = \begin{cases} 5T - 0.1 & T \le 0 \\ 0.5T - 0.1 & 0 < T \le 4 \\ -3T + 13.9 & T > 4 \end{cases} \tag{5.1}$$

The equations capture the emotional state of each of the participants as it is determined by participant's own state when uninfluenced (b); participant's own inertia (m), which is dependent on one's own previous emotional state; and the influence each of the partners has on the other one. The functional form of the influence functions in the model is based on published empirical results and researchers' experience by considering what would be a typical emotional reaction of a client and therapist to a display of negative, neutral, or positive emotion by their respective partner of interaction.

Based on the model with the above-discussed parameters and influence functions, Liebovitch and Peluso (Liebovitch et al. 2011; Peluso et al. 2012) conducted simulations. They examined how different circumstances (e.g., different initial emotional state of the participants) affect the process of psychotherapy and its

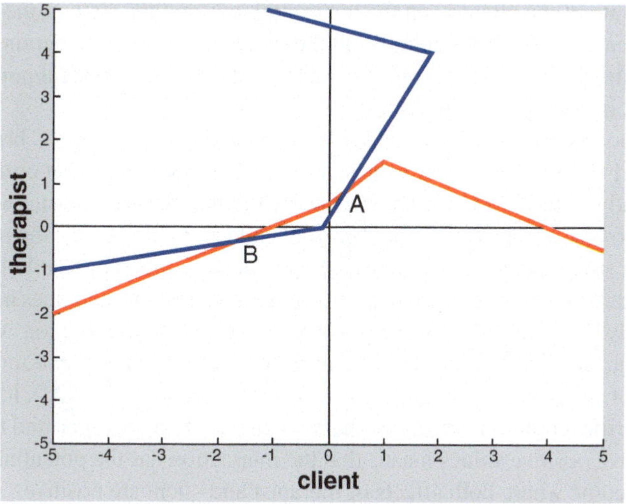

Fig. 5.1 The intersection of the null clines for a model with the same parameter values for the therapist and client, $m = -1$, $b = 0$, $c = 1$. Two critical points A and B are visible

possible outcomes. To get insight into the dynamics of the interaction between therapist and client, they utilized graphical representation of the systems evolution in time. Two types of such helpful visualization tools, namely, graphs of null clines and phase space portraits, will be discussed in the subsequent paragraphs. These graphs will also be later used to visualize the simulation results when the new component, the relationship repair attempt, is included in the model.

In these graphs the emotional state of the client is depicted on the horizontal axis, and the state of the therapist is depicted on the vertical axis (Fig. 5.1). For the purpose of illustration, all but one discussed scenarios depict the emotional state of the client and the therapist within the range of −5 and 5. Negative numbers correspond to negative emotional states, and positive numbers correspond to positive emotional affect. The higher the absolute value of the number, the greater the intensity of a given affective state. One scenario depicts values of the emotional states in the range of −10 and 10, because for this particular scenario, an interesting dynamics corresponds to a situation where the client's emotional state equals to −7 (see Fig. 5.5).

The null-cline graph depicted on Fig. 5.1 can be explained in the following way. The emotional state of both, the therapist (T) and client (C), starts from some initial values, and both evolve in time. To determine the outcome of the therapy for both the therapist and client, steady states have to be identified. If they exist, they must occur where the values of T and C are no longer changing in time, namely, where the derivatives $dT/dt = 0$ and $dC/dt = 0$. The special values of T and C where this occurs are called the critical points. The equations for $dT/dt = 0$ and $dC/dt = 0$ are each called the null clines. Since the critical points satisfy the equations of both null clines, they lie at the intersections of the two null clines. To see the values of these critical points graphically as the intersections of the null clines, a plot is created where the state of the client is on the horizontal axis and the state of the therapist is on the vertical axis. To fully understand the dynamics of the interaction around the critical points, one more step is necessary. The nature of each critical point has to be determined, in particular their stability.

Critical points can be stable or unstable. If small changes in T and C bring both the therapist and client back to their values at the critical point, then that critical point is a stable steady state. This is a point of equilibrium that the relationship would gravitate toward (or be attracted to). It is therefore called an attractor for the relationship. However, if small changes in T and C always push the therapist and client further away from the critical point, then that critical point is unstable and does not represent a final steady state. In the example presented in Fig. 5.1, intersection A is a stable critical point, and intersection B is an unstable critical point (for a more detailed discussion of linear stability analysis, please see Gottman et al. 2002). Identification of a stable critical point in the space where the affect of the therapist and the affect of the client have positive values means that the interaction has the potential to end in a positive outcome where both affects of therapist and client are positive.

The system dynamics around the critical points is further best visualized with a phase space portrait, which depicts the temporal trajectory of the system from various initial conditions (see Fig. 5.2). It shows not only the location of the critical points but also possible trajectories leading to a given outcome. Each point in this phase space corresponds to one value T for the affective state of the therapist on the vertical axis and one value C for the affective state of the client on the horizontal axis. The therapist and client each start with an initial affective state called their initial condition. Their initial condition, having one value for T for the therapist and

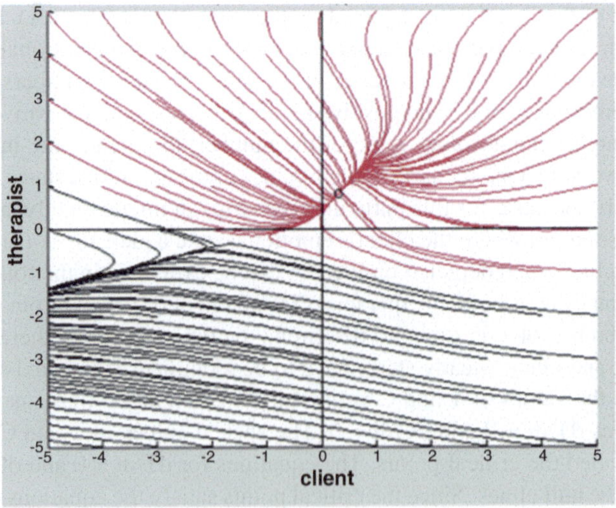

Fig. 5.2 The phase space portrait for a model with parameter values for the therapist and client as follows: $m = -1$, $b = 0$, $c = 1$

one value for C for the client, is represented by one point in the plot. The dynamics of the therapeutic relationship can then be traced by following the path of this point, called its trajectory, as it moves through the plot toward or away from a critical point.

Liebovitch and Peluso (Liebovitch et al. 2011; Peluso et al. 2012) showed that this mathematical model captures important features of the therapeutic relationship. For example, they showed that the affective exchange between the therapist and client may cause the clients to go through emotional ups and downs and eventually let their emotional state to stabilize at a critical point. For a comprehensive overview of the simulation results, readers are encouraged to see Chap. 4 in this volume.

5.5 Including Repair Terms in the Mathematical Model of the Therapeutic Relationship

Gottman and his associates found that adding repair terms to their model greatly enhanced the predictive value of the mathematical model of marriage (Gottman et al. 2002). A repair is the attempt to move a relationship away from a negative trajectory. In the context of a therapeutic relationship, the negative trajectory corresponds to a potential therapeutic rupture. This section presents how the addition of the repair term to the original model of the therapist-client relationship affects the trajectory of the interaction through an analysis of the phase space portraits, the steady-state location, and stability.

5.5.1 Modeling the Repair Term

In order to introduce the repair term to the model, a new component (or parameter) is added to the model equation. This new parameter should capture the fact that interactants are responsive to overly negative state of the partner and try to prevent the partner from becoming too negative. They can do so by changing their own state into a more positive one increasing the chances that the partner in turn will respond with more positivity. The repair component in the model will therefore represent the change to a participant's own score in response to the partner's score becoming more negative than a certain threshold.

$$R_T(C) = s_T \frac{|C - k_C| - (C - k_C)}{1 - (C - k_C)} \tag{5.2}$$

Equation 5.2 presents an exact mathematical expression of the repair term for the therapist repairing (R_T) if client's affect (C) becomes more negative than a certain threshold (k_C). The repair term is characterized by two parameters, the strength (s) and threshold (k). The strength of repair indicates how much the repairing therapist changes his or her own score in response to the negativity of the client. A weak repair attempt will not increase the therapist's score as much as a strong repair attempt. As a consequence, there is a smaller chance for the weak repair attempt in comparison to a strong repair attempt that the intervention will change the negative affect of the client. The repair threshold (k) refers to the value of the client's score that will trigger the repair attempt. In the context of the therapist repairing, the repair threshold is denoted by the client's score that is the most negative score not met with a repair attempt by the therapist. In other words, this is the most negative affect the therapist will not react to by trying to change it through repairing the relationship. As soon as the client's affect becomes more negative than the threshold, the therapist starts repairing. When the repair attempt occurs in response to even a fairly mildly negative affect, we can term it an early repair. If the repair happens only in response to a very negative affect, it is a late repair. For the purpose of clarity of the discussion, the effects of an early repair will be discussed separately from the effects of a late repair attempt. The repair term can be introduced on both sides, in which case it means that both the therapist and client will be sensitive to their partner's overly negative affect or it may just be one-sided, for example, with only the therapist trying to prevent the relationship rupture.

To investigate the effects of repair attempts on the dynamics of interaction, a set of computer simulations were conducted. The results of those simulations are presented and discussed below.

5.5.2 *Simulation of the Model*

5.5.2.1 The Effect of Therapist Repairing

When the therapist reacts to the negative affect of the client by repairing, a new steady-state C appears in the region where both the therapist and client experience negative affects. We call it a new steady state in comparison to the original model that had only two steady states, A and B, as discussed in section 5.3 earlier. The exact location of that new steady state corresponds to the value of the threshold parameter (see Figs. 5.3 and 5.4).

Conceptually, within the context of a therapeutic relationship, a rupture may be the result of a therapist noting that the client has become too negative. The therapist repairs this by making his or her own affect more positive. This change in therapist affect has the potential to make the client more positive through the regular influence function. The result is that instead of going to minus infinity as before, the interaction has a chance to end up in a negative, yet stable state (denoted by C on Fig. 5.3).

The Effect of Therapist Early Repairing Even the weakest early repair attempt ($s = 1$) results in a creation of two new critical points, one of which is a steady state (attractor) in the negative-negative region ($c = -2$, $t = -0.38$) and the other is a saddle point situated in the negative-negative space ($c = -7$, $t = -1.4$) (see Fig. 5.5).

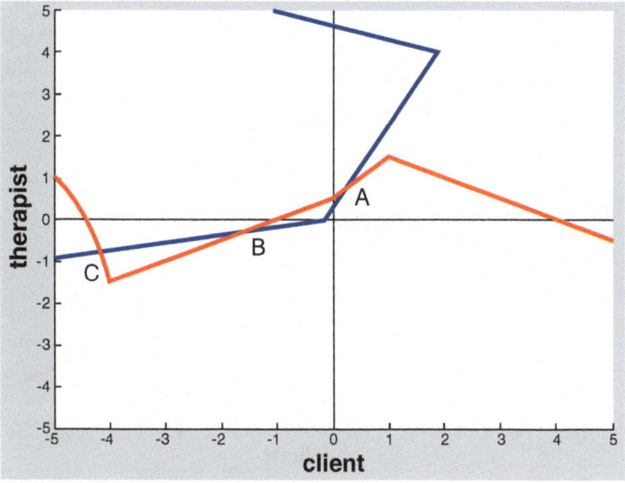

Fig. 5.3 The null clines for the model with repair term

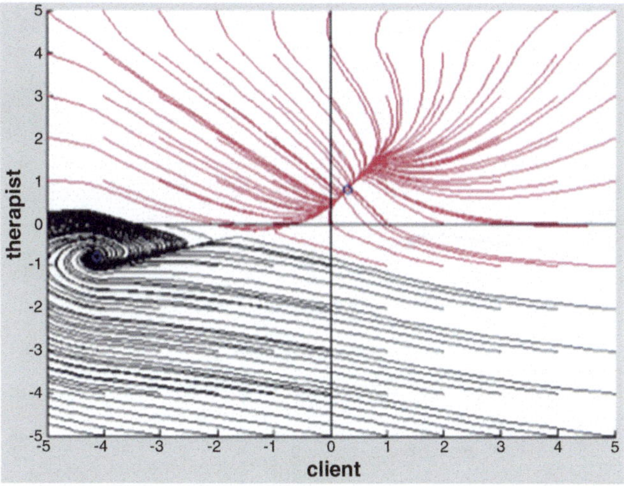

Fig. 5.4 The phase space portrait of the model with repair term

We next analyzed changes in null clines and phase space portraits at increasing values of the parameter reflecting the strength of the repair attempt, and we noticed the following pattern. Increasing the strength of an early repair attempt does not affect the location of any of the two stable points identified also in the original model. Interestingly, as the strength of the repair attempt increases, the number of trajectories that lead to the positive steady state (basin of attraction) increases at the expense of those that lead to the negative steady state. Figure 5.6 and Fig. 5.7 show two phase portraits that demonstrate the effect of increasing the strength of an early

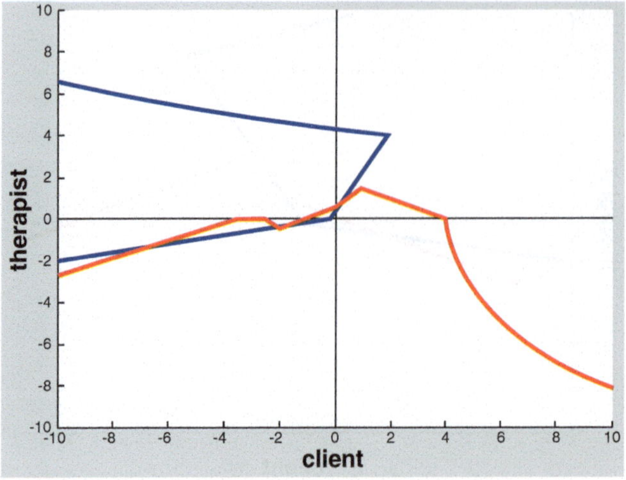

Fig. 5.5 The null clines for the model with a weak early repair attempt by the therapist

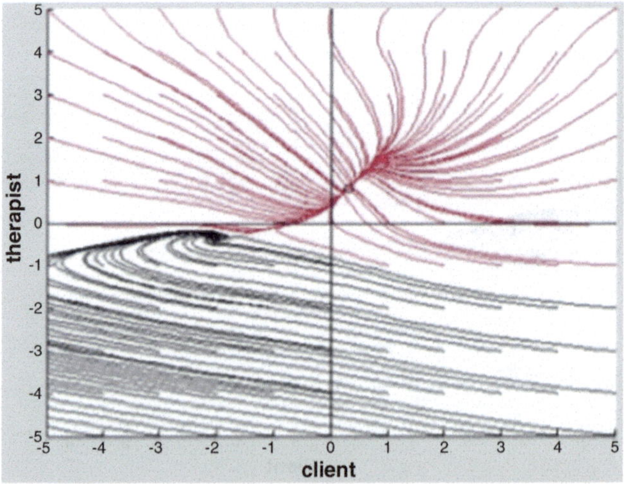

Fig. 5.6 Phase portraits showing the effect of weak early repair attempt by therapist

repair attempt ($k = -2$) by the therapist. The increased size of the basin of attraction for the positive steady state is denoted by more trajectories leading to the steady state. For example, scenarios where the client started with positive affect and the therapist started with negative affect would lead to a negative state for both of them in the situation with a weak repair (Fig. 5.6). However, when the therapist attempts repair more strongly, it allows for such configurations to lead to an outcome where both therapist and client have positive affect (Fig. 5.7).

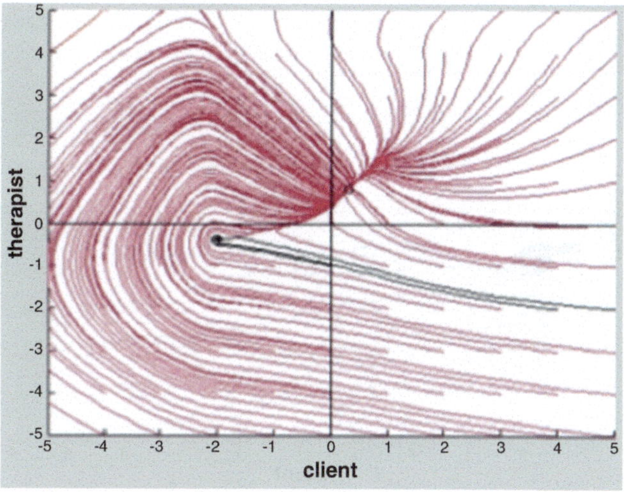

Fig. 5.7 Phase portraits showing the effect of strong early repair attempt by the therapist

This model demonstrates how clinicians who recognize early negativity in the client's affect can avoid a negative outcome by effectively repairing the therapeutic rupture or preventing it altogether.

The Effect of Therapist Late Repairing While an early intervention is preferable, even therapists who respond late (e.g., after a rupture has begun) can still make a positive impact. Figures 5.8 and 5.9 show two phase portraits that demonstrate the effect of a late repair attempt ($k = -4$) by the therapist. Again, even a weak late repair ($s = 2$) results in creation of a new steady state (attractor) at the following affect values for the client and therapist, respectively, $c = -4.3$, $t = -0.83$.

As the strength of a late repair attempt increases, the basin of attraction for the positive steady state increases in size at the expense of the size of the basin of attraction for the negative steady state (Figure 5.7b).

The difference between what happens when the repair attempt occurs early versus late is apparent when comparing Fig. 5.6b with Figure 5.7b. In the latter case, the basin of attraction for the negative steady state remains larger than in the case of an early attempt. In the early repair model, clients are more likely to be drawn into the positive attractor. Therapeutically, the clinician may express his honest dissatisfaction with the way that the therapy is progressing with the client. The therapist may take ownership that they have "missed" something or misunderstood the client. Alternatively, the therapist may utilize a paradoxical intervention that also has the effect of repairing the relationship (Mozdzierz et al. 2014).

Variations of Repairing by the Therapist and its Effect on the Client One of the unique features of the dynamical systems model is that small variations in the parameters of the equations can create significant effects on the model itself. This

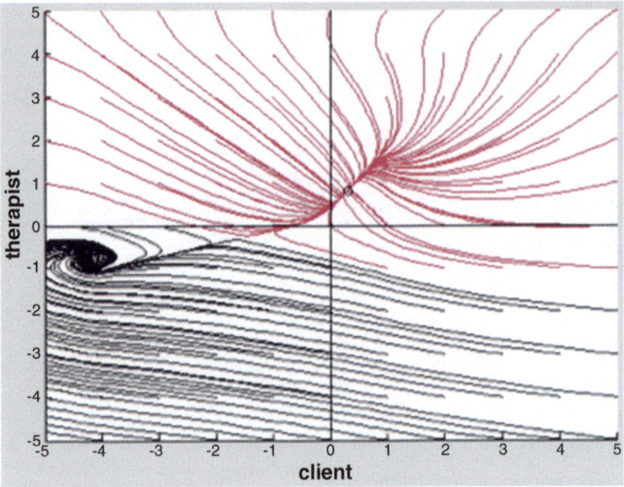

Fig. 5.8 Phase portraits showing the effect of weak late repair attempt by the therapist

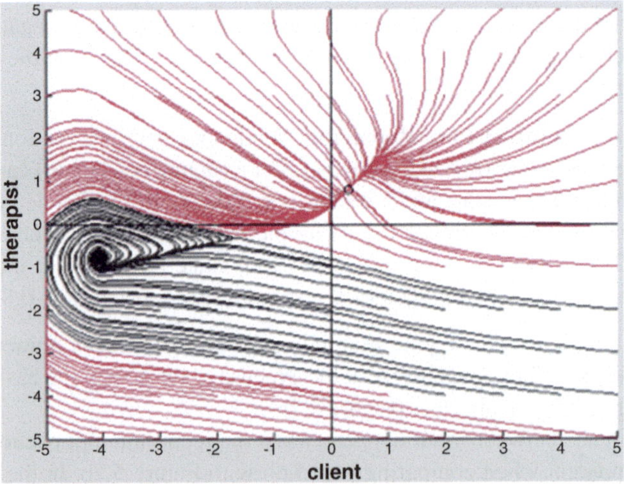

Fig. 5.9 Phase portraits showing the effect of strong late repair attempt by the therapist

can be represented in the phase portraits and has been shown in some of the previous comparisons for different strength values. It was also shown that by varying the threshold for the repair (which is the moment when the therapist starts repairing) at the same level of strength, clinicians can see how the relationship (viz., the basin of attraction for the steady states) will change. When the client's state reaches the value of this repair threshold, the therapist starts repairing by becoming more positive, and this in turn has the potential to elicit a more positive response on the side of the client. Figs. 5.10, 5.11, 5.12 and 5.13 show how changing (increasing) repair threshold (k) affects the phase space portrait for a fixed value of repair strength ($s = 4$).

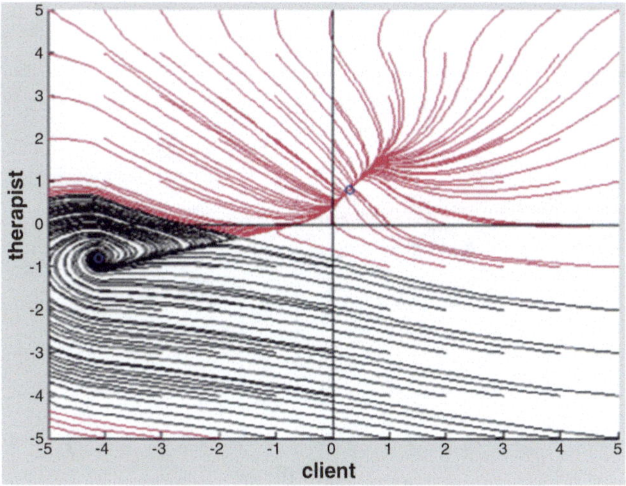

Fig. 5.10 Phase portraits showing the effect of increasing repair threshold kr = −4

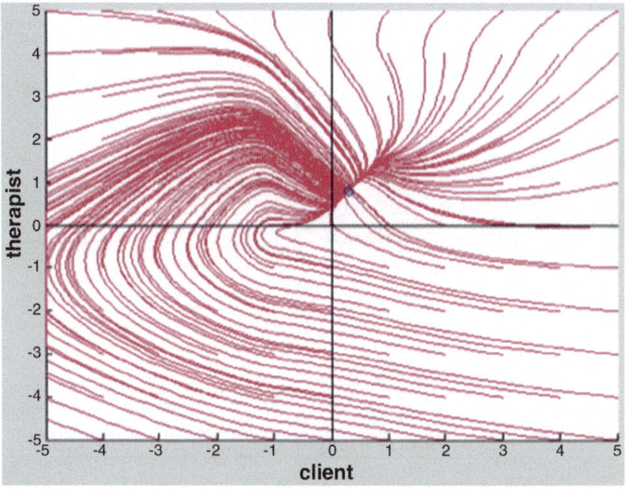

Fig. 5.11 Phase portraits showing the effect of increasing repair threshold kr = −3

Varying the threshold coefficient shows that the location of the new steady state in the negative-negative region closely corresponds to the value of the threshold set for repair attempt. As the threshold changes from −4 to −2, the negative steady state moves accordingly along the client axis, and the basin of attraction for this negative steady state shrinks in size with more trajectories leading to the positive steady state, which is depicted in Figs. 5.10, 5.11, 5.12 and 5.13. Table 5.1 shows coordinates (where x = client score, y = therapist score) of the negative steady states for a given value of repair threshold.

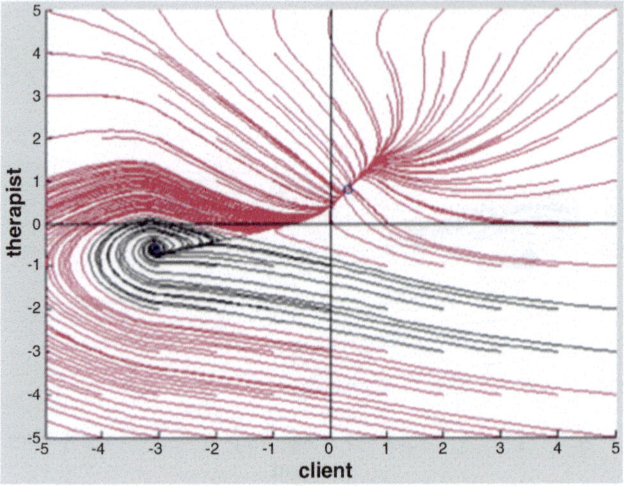

Fig. 5.12 Phase portraits showing the effect of increasing repair threshold kr = −2

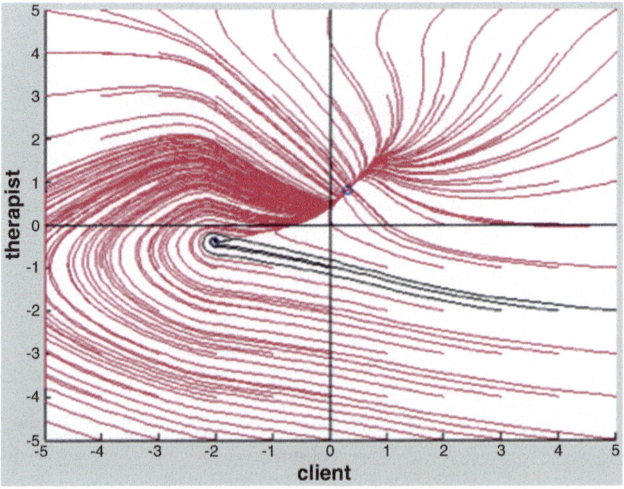

Fig. 5.13 Phase portraits showing the effect of repair threshold kr = −1

Finally, when the threshold is equal to −1, the negative steady state ceases to exist, and only one positive steady state remains (see Fig. 5.13). It is important to note how this last scenario is different from the scenario offered by the original model without the repair term that also had only one positive steady state (compare Fig. 5.2 and Fig. 5.13). Here, all starting points lie within the basin of attraction of the positive steady state suggesting that a very early repair attempt ($k = -1$) should always result in an interaction ending up in a positive steady state.

Table 5.1 Negative steady-state location for increasing values of repair threshold

Kr (repair threshold)	−4	−3	−2
c (client's score)	−4.1	−3.1	−2.0
t (therapist's score)	−0.8	−0.59	−0.38

5.5.2.2 The Effect of Client Repairing

Interestingly, even though both members of the dyad should be able to effect change, according to the psychotherapy model of Liebovitch and Peluso (Liebovitch et al. 2011; Peluso et al. 2012), the magnitude of the effect of the client attempting a repair to the relationship is not equivalent to the magnitude of the effect of the therapist attempting a repair. This is demonstrated in the simulations to the model when the client attempts a repair. For example, if the early repair is defined as the threshold coefficient set to −2 (per analogy to the exploration presented for the therapist's responses), then increasing the strength of the repair term has no effect on the number and location of the steady states. If, however, the client would react even to the fairly slight negative affect of the therapist ($k = -1$), then with a strong tendency to repair ($s > 5$), a new steady state appears in the negative-negative space (Fig. 5.14).

While this new steady state is located in the negative-negative space, it is only slightly different than the original model (see Fig. 5.2), and the basin for the positive attractor is not significantly changed (as it is when the therapist is the person attempting the repair).

The Effect of Client Late Repairing Introduction of a late repair term ($k = -4$), no matter how strong it is, has no effect on the location and number of steady states in the system. Therefore, the phase space portrait was not reprinted here.

Variations of Repairing by the Client and its Effect on the Therapist Changing the threshold of the repair does not affect the dynamics of the system. Only if the repair response by the client is strong ($s > 5$), there will be another negative attractor formed in the negative-negative space when the threshold is equal to −1 (see Fig. 5.14).

5.5.3 *Simulation Result Summary*

5.5.3.1 Therapist Repairing

When the therapist reacts to the negative affect of the client by repairing, that is, by becoming more positive, which in turn has the potential to elicit a more positive response on the side of the client, a new steady state appears in the region of negative-negative affect. As a result, the interaction may end up either in the positive steady state or in the negative steady state. It is important to note that even if the interaction is likely to end up in the negative steady state, it is still a better outcome

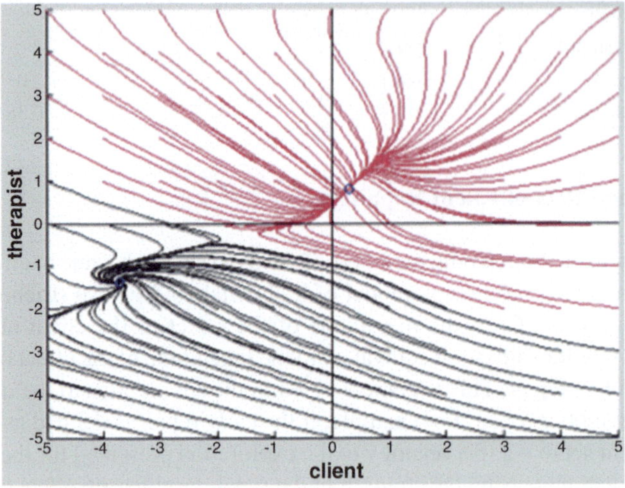

Fig. 5.14 Phase portrait when client is strongly repairing ($s = 6$) in response to even slightly negative affect ($k = -1$) of the therapist

for the therapy than scenarios predicted by a model with no repair attempts, where the therapy could (under certain conditions determined by the model parameters and the starting conditions of the participants) either end up in the positive steady state or end with emotional trajectories of both partners leading to minus infinity. In the scenarios offered by the original model, the characteristics of the interaction partners (state when alone and inertia) and a quality capturing their mutual influence (coupling) would either allow them to achieve a favorable outcome or not, while the new scenario that involves repairing interventions on the side of the therapist allows the therapist to prevent the client from falling below a threshold that the therapist deems to be unfavorable for the client.

Therapist Repairing Early The analysis of the threshold of the repair leads us to the conclusion that in order to increase the probability that the interaction will end up in the positive steady state, the therapist should start repairing early, that is, before the client's state becomes very negative. The part of simulation where the threshold was varied shows that an early repair attempt by the therapist might be a more effective intervention (as compared to late repair or no repair at all) leading to more positive final affective state achieved by the client. The final affective state of the client corresponds closely to the repair threshold suggesting that a repair intervention of the therapist is an effective way of preventing the client from falling into too negative affective states that the therapist considers detrimental to the therapy process.

Therapist Repairing Strongly In addition, a stronger repair attempt, which translates to a more pronounced shift toward a positive state by the therapist in response to the client's negativity, leads to a situation where there are smaller chances that the interaction will end in the negative stable state, but it will rather gravitate toward the positive stable state. Hence, in order to avoid a negative outcome, therapists must work hard to identify and repair the ruptures in the therapeutic relationship as vigorously as they can.

5.5.3.2 Client Repairing

Exploration of the repair attempts by the client shows that when the client is very responsive and therefore strongly reacts even to the fairly slight negative affect of the therapist, only then a new steady state is visible in the negative-negative space meaning that the interaction has a chance to end up in a negative, yet a better state than the minus infinity predicted by the model without the repair component.

5.5.3.3 Implications of Simulation Results

The results of the simulation of the model with repair term are valuable in two ways. First, they match observational and experimental data on psychotherapy relationship. The results also inspire new hypotheses that can be used to verify the predictions of the model empirically.

Results that Match Previous Observations and Experiments Some of the results of the model match previous observations about the dynamics of psychotherapeutic relationship. According to Safran et al. (2011), attending to and repairing therapeutic ruptures are crucial in maintaining a successful therapeutic relationship. A repair attempt that is successful can greatly enhance any successes in therapy. In the simulations above, this success is modeled and graphically represented. It is also consistent with Gottman et al.'s (2002) assertions regarding the success of marital dyads.

At the same time, avoiding a rupture altogether is a superior strategy. In particular, Duncan et al. (2010) advocate for a "feedback-informed therapy" (FIT) in which the therapist solicits client input about the course of a given session or therapy altogether. This feedback forms the basis of changes in strategy or approach in the therapist—essentially a repair to client dissatisfaction (negativity) in the relationship. In our modeling, the therapist's openness to influence is a factor that can be adjusted to see the magnitude of impact of the feedback and response to it (as visualized on Figs. 5.5 and 5.8). Miller and his colleagues (Duncan et al. 2010; Miller et al. 2010; Wampold 2010) have demonstrated the effectiveness of this approach.

Predictions of the Model Interestingly, our modeling presents three important clinical predictions. (1) Repairs to the therapeutic relationship can create a new steady state. (2) The timing and strength of the repair are important in determining the success of the repair. (3). The client's repairing does not seem to affect the dynamics of the relationship as dramatically as the therapist's repairing. This is consistent with the clinical literature (Duncan et al. 2010; Norcross 2010; Safran et al. 2011).

The results of these simulations provide also new testable predictions on how the dynamics of the psychotherapeutic relationship and its outcomes depend on the strategies of the actors. Below we outline, how the model predictions can, and in fact will be, empirically verified.

5.6 Future Directions

The effects of the repair attempts by the therapist can be empirically investigated through observation of the therapist and client affect during a psychotherapy session. This follows from Duncan et al.'s (2010) advice that "Instead of repeatedly testing the overall efficacy of whole treatment packages (i.e., models), attention is turned to the moment-by-moment, encounter-to-encounter processes associated with effective psychotherapy" (p. 425), as well as the following recommendation by Norcross and Wampold's (2011):

> Researchers are encouraged to progress beyond correlational designs that associate the frequency of relationship behaviors with patient outcomes to methodologies capable of examining the complex associations among patient qualities, clinician behaviors, and treatment outcome. Of particular importance is disentangling the patient contribution and the therapist contribution to relationship elements and ultimately outcome... Researchers are encouraged to examine the specific mediators and moderators of the links between the relationship elements and treatment outcome (p. 425).

A procedure adapted from Gottman's specific affect (SPAFF) coding system captures the moment-to-moment changes in the affect of the participants based on their facial expressions, vocal tones, as well as speech content (Gottman et al. 1996; see also Chaps. 3 and 4 in this volume). Specifically, during the interaction the five positive codes (interest, validation, affection, humor, and joy), ten negative affect codes (disgust, contempt, belligerence, domineering, anger, fear/tension, defensiveness, whining, sadness, stonewalling) and a neutral affect are identified from video recordings. The model assumes that a negative affect on the side of the client can be followed by the repair response where the therapist displays more positive affect in order to elicit a more positive response from the client. We see two ways of verifying if the repair attempts occurred in a given session. First, the repair attempts should be visible in the therapist's affect scores at certain (low) values of the client's affect scores. In this case, the repair threshold would be estimated based on the client scores that are followed by an increase in positivity in therapist's scores. Clinically, repairs create a shift in meaning in relationships (Gottman 2011). Successful repairs can radically alter a client's perception of the therapy, whether it is accomplished before or after a rupture has occurred. Observational data can provide a metric for determining the process, as well as the result of these repairs relative to the modeling reported above. The second approach involves a post-recording session with the therapists that while watching the video recording would indicate where they in fact attempted a repair intervention. We expect the SPAFF codes to reflect these repairing interventions.

The affect coding method will also be useful in investigating the relationship between the affect dynamics and psychotherapy outcomes. Specific patterns of affect dynamics can be correlated with treatment success measures such as retention, premature termination, problem-specific outcome measures such as Beck Depression Inventory-II (BDI-II), as well as subjective measures of psychotherapeutic relationship, for example, the Real Relationship Inventory that captures the client's perceptions of the genuineness and realism of the therapist or the Working Alliance Inventory that assesses the therapist's and client's judgment of the relative strength and quality of the therapeutic alliance.

The present simulation work produced a model that extends Liebovitch et al. (2011) and Peluso et al.'s (2012) dynamical systems model for the therapeutic relationship by adding a repair term to the model parameters similar to Gottman et al. (2002). As presented, in this simulation work, we followed the iterative procedure of introducing variations into the model to explore how different interventions may affect the outcomes. We hope that this detailed account of our simulation work shows that mathematics, as expressed by Estes, can serve as:

> (…) our principal vehicle for the flights of imagination that smooth our experiences and extract from varying contexts the relationships that would hold among events under idealized noise-free conditions. These abstract representations are not necessarily closely descriptive of data obtained in real noise-filled environments, but in the course of interactions with data they generate the reorganizations of our ideas that constitute new theory. (Estes 1975, p. 279)

This new model suggests that therapies where the therapist uses repair as a means of eliciting a more positive affect from the client might have beneficial influence on the psychotherapy outcomes; therefore, those therapies that display an affect dynamics with repair interventions are expected to correlate positively with quality of the psychotherapeutic relationship and effectiveness of psychotherapy (as indicated by the problem-specific measures) and be inversely related to retention rates and premature termination. It is hoped that subsequent analysis of empirical data from actual psychotherapy sessions will allow for testing and further refinement of this model.

References

Bordin, E. S. (1994). Theory and research on the therapeutic working alliance: New directions. In A. O. Horvath & L. S. Greenberg (Eds.), *The working alliance: Theory, research and practice*. New York: Wiley.

Coutinho, J., Ribeiro, E., Hill, C., & Safran, J. (2011). Therapists' and clients' experiences of alliance ruptures: A qualitative study. *Psychotherapy Research, 21*(5), 525–540. https://doi.org/10.1080/10503307.2011.587469

Duncan, B. L., Miller, S. D., Wampold, B. E., & Hubble, M. A. (Eds.). (2010). *The heart and soul of change: Delivering what works in therapy* (2nd ed.). Washington DC: American Psychological Association.

Estes, W. K. (1975). Some targets for mathematical psychology. *Journal of Mathematical Psychology, 12*(3), 263–282. https://doi.org/10.1016/0022-2496(75)90025-5

Gottman, J. (2011). *The science of trust: Emotional attunement for couples*. New York: W. W. Norton.

Gottman, J. M., McCoy, K., Coan, J., & Collier, H. (1996). The specific affect coding system (SPAFF). In J. M. Gottman (Ed.), *What predicts divorce? The measures*. Mahwah: Lawrence Erlbaum.

Gottman, J., Murray, J., Swanson, C., Tyson, R., & Swanson, K. (2002). *The mathematics of marriage: Dynamic nonlinear models*. Cambridge, MA: MIT Press.

Lambert, M. J. (1992). Psychotherapy outcome research: Implications for integrative and eclectic therapists. In J. C. Norcross & M. R. Goldfried (Eds.), *Handbook of psychotherapy integration* (pp. 9–129). New York: Basic Books.

Lambert, M. J., & Barley, D. E. (2002). Research summary on the therapeutic relationship and psychotherapy outcomes. In J. C. Norcross (Ed.), *Psychotherapy relationships that work: Therapist contributions and responsiveness to patient needs* (pp. 17–32). New York: Oxford University Press.

Liebovitch, L. S., Naudot, V., Vallacher, R. R., Nowak, A., Bui- Wrzosinska, L., & Coleman, P. (2008). Dynamics of two-actor cooperation-competition conflict models. *Physica A, 387*, 6360–6378. https://doi.org/10.1016/j.physa.2008.07.020

Liebovitch, L. S., Peluso, P. R., Norman, M. D., Su, J., & Gottman, J. M. (2011). Mathematical model of the dynamics of psychotherapy. *Cognitive Neurodynamics, 5*, 265–275. https://doi.org/10.1007/s11571-011-9157-x

Miller, S. D., Hubble, M. A., Duncan, B. L., & Wampold, B. E. (2010). Delivering what works. In B. L. Duncan, S. D. Miller, B. E. Wampold, & M. A. Hubble (Eds.), *The heart and soul of change: Delivering what works in therapy* (2nd ed., pp. 421–429). Washington, DC: American Psychological Association.

Mozdzierz, G., Peluso, P. R., & Lisiecki, J. (2014). *Principles of counseling and psychotherapy: Learning the essential domains and nonlinear thinking of master practitioners*. New York: Routledge.

National Institutes of Mental Health. (2010). *The numbers count: Mental disorders in America*. Retrieved from https://www.nimh.nih.gov/health/publications/the-numbers-count-mental-disorders-in-america/index.shtml.

Norcross, J. C. (2010). The therapeutic relationship. In B. L. Duncan, S. D. Miller, B. E. Wampold, & M. A. Hubble (Eds.), *The heart and soul of change: Delivering what works in therapy* (2nd ed., pp. 113–142). Washington, DC: American Psychological Association.

Norcross, J. C., & Wampold, B. E. (2011). Evidence-based therapy relationship: Research conclusions and clinical practices. *Psychotherapy, 48*, 98–102. https://doi.org/10.1037/a0022161

Peluso, P. R., Liebovitch, L. S., Gottman, J. M., Norman, M. D., & Su, J. (2012). A mathematical model of psychotherapy: An investigation using dynamic non-linear equations to model the therapeutic relationship. *Psychotherapy Research, 22*, 40–55. https://doi.org/10.1080/10503307.2011.622314

Safran, J. D., & Muran, J. C. (2000). *Negotiating the therapeutic alliance: A relational treatment guide*. New York: Guilford Press.

Safran, J. D., Muran, J. C., Samstang, L. W., & Stevens, C. (2002). Repairing alliance ruptures. In J. C. Norcross (Ed.), *Psychotherapy relationships that work: Therapist contributions and responsiveness to patient needs* (pp. 235–254). New York: Oxford University Press.

Safran, J. D., Muran, J. C., & Eubanks-Carter, C. (2011). Repairing alliance ruptures. *Psychotherapy, 48*(1), 80–87. https://doi.org/10.1037/0033-3204.38.4.406

Wampold, B. E. (2010). The research evidence for common factors models: A historically situated perspective. In B. L. Duncan, S. D. Miller, B. E. Wampold, & M. A. Hubble (Eds.), *The heart and soul of change: Delivering what works in therapy* (2nd ed., pp. 49–81). Washington DC: American Psychological Association.

Chapter 6
Mathematical Models as Tools for Understanding the Dynamics of Cooperation and Conflict

Jay L. Michaels

6.1 Why Models

Few forms of social interaction have as powerful an influence on human behavior as cooperation and conflict. That destructive conflicts and inspiring acts of cooperation appear prominent in forms of human expression tracing to the roots of civilization (e.g., a core theme in the Epic of Gilgamesh, c. 2100 BCE) implies these phenomena have a uniquely important role in people's lives. Yet, for thousands of years, answers as to why some conflicts seem stubbornly resistant to resolution—and how new cooperation can suddenly emerge between former adversaries—remain elusive.

Part of the challenge in understanding these social phenomena is that conflict and cooperation are among the most complex social behaviors. The evolution of cooperation and conflict is linked to factors intrinsic to the people involved, the context in which the interactions take place, a myriad of psychological processes, the history of the social situation, and the expectancy of future outcomes (e.g., Coleman 2011; Kriesberg 1980; Vallacher et al. 2010a). While valuable insight may be gleaned by examining any of these factors individually, Pruitt and Olczak (1995) note that advancing understanding of cooperation and conflict requires forming an understanding of the "synergy" in an interaction, where the "interaction among several causal forces" yield a "joint impact [that] is greater than or different from the sum of their individual components" (p. 85). Thus, understanding cooperation and conflict requires recognition of these phenomena as related processes emerging from multiple situational as well as inter- and intrapersonal components that interact over time (e.g., Deutsch et al. 2006). The idea that human interactions are products of many mutually influential, interactive factors bound to an evolving situation

J. L. Michaels (✉)
Department of Psychology, University of South Florida Sarasota-Manatee, Sarasota, FL, USA
e-mail: jaymichaels@sar.usf.edu

U. Strawinska-Zanko, L. S. Liebovitch (eds.), *Mathematical Modeling of Social Relationships*, Computational Social Sciences,
https://doi.org/10.1007/978-3-319-76765-9_6

is evident across behavioral science paradigms. For example, complexity science is now integral to most, if not all, areas of science and transcends traditional disciplinary bounds (e.g., Gleick 1987, Johnson 2001). The idea that society and group interactions operate as ever-changing dynamic systems is integral to contemporary psychology (e.g., Guastello et al. 2009; Vallacher et al. 2015; Westaby et al. 2014). These ideas are hardly limited to basic research and theory; for example, as Burns (2007) noted of systemic action research, an approach to finding new solutions to challenging social problems, "complex issues cannot be adequately comprehended in isolation from the wider system of which they are part" (p. 1). While the idea that cooperation and conflict function as complex, dynamic systems is doubtlessly accepted in contemporary behavioral science, an integral challenge remains. Making sense of multiple factors that frequently exert nonlinear influences on one another is challenging.

Fortunately, advances in computing and concepts adopted from complexity science (e.g., see Nowak et al. 1994; Strogatz 1994; Sun 2008) have inspired innovative researchers to develop insightful models of cooperation and conflict. Before examining specific models in greater detail, it is worthwhile to consider how models are important, if not central, to contemporary science. Foremost, models fundamentally differ from theories in that social science theories serve to coherently integrate and interrelate knowledge about behaviors. In contrast, *models* are specific descriptions of the core processes that influence and give rise to behaviors (cf. Heiman 2001). Models are not necessarily mathematical (e.g., various models of self-control, e.g., Inzlicht and Schmeichel (2012); affect-as-information and related models, e.g., Isbell et al. (2013), Schwarz and Clore (2007)), though the precise emphasis of specific factors and their interrelatedness naturally affords greater ease at translating the ideas into mathematical expressions.

Why are mathematically expressed models of value in science? Foremost, as Richardson (1960) elaborated, "to have to translate one's verbal statements into mathematical formulae compels one carefully to scrutinize the ideas therein expressed," which clarifies consequences and "absurd implications" inherent in the model (p. xvii). The clarity a parsimonious model provides about a complex phenomenon, such as cooperation and conflict, facilitates precise predictions that are useful for formulating empirically testable hypotheses (see Nowak 2004). Indeed, models are pertinent for the scientific process as a whole, as Hunt (2007) emphasized, "science does *not* advance by proving theories correct; it advances by developing models that are progressively more and more accurate" (p. 14). It is worth emphasizing that models are not superior to theories. Rather, models are a vital tool for translating theories (especially complex theories) into precise concepts and relationships that may be operationalized for empirical research. Of course, models have limitations, as do all tools applied in science. Richardson (1960) again provided insight, cautioning that "mathematical expressions have, however, their special tendencies to pervert thought: the definiteness may be spurious, existing in the equations but not in the phenomena to be described; and the brevity may be due to the omission of the more important things simply because they cannot be mathematized" (p. xviii). Developing models that accurately capture behavior also proves challenging as the problems of interest are often difficult, have greater depth than may be first anticipated, and typically transcend disciplinary bounds (see Sloman 2008). While developing models may thus take time and add iterations to the research cycle—as models are tested against evidence, refined, and retested—there is little doubt that computational models will continue to provide important, sometimes nonobvious insight into complex problems (cf. Liebovitch et al. 2010). The insights gained from models may thus prove informative for theory building and refinement in addition to practices in conflict resolution and negotiation.

6.1.1 Models and the Scientific Process

As with empirical research first requiring the phenomena of interest be bounded for field or lab study, mathematical modeling begins with precise definition. There are different approaches to specifying which variables related to the theoretical construct are included in a model. Familiar to many in the social sciences is a technique to refine a theoretical construct into specific variables that can be translated into operational definitions leading to specific forms of measurement. The data obtained from such measurement is then analyzed using inferential statistics, such as regression modeling, where variables achieving statistical significance are included in a final model, and those that fail to reach significance are excluded. While this approach has long proven valuable in advancing understanding of many social phenomena, there is an alternative less common in the social sciences.

In many of the physical sciences, especially, models are developed as a means to formalize a theory, which then leads to empirical research that tests and refines the model (cf. Hunt 2007). In recent years, this progression has been evident within physics, where mathematical models predicted expected properties of the Higgs boson, a unique particle considered central in particle physics (e.g., Gunion et al. 1990). These models led to precise empirical tests using the European Organization for Nuclear Research's particle accelerators to discover evidence for the existence of the Higgs boson (e.g., Chatrchyan et al. 2013). Other classic examples of how models led to important scientific discoveries and advances include planetary laws, being used to predict and then discover Neptune, and Lorenz's equations, proving fundamental for developing atmospheric models used daily to predict weather.

Formulating models from theory is not an easy task, nor is there a rigid logic as to how mathematical models interface with the broader science research process. Models are perhaps best considered one step of scientific inquiry. As Nowak (2004) reflected:

> The development of a theory happens through repeated iteration of computer simulations that investigate the properties of the theoretical model and empirical investigations that test the model's assumptions and predictions. Without the empirical component, the value of the theory is greatly diminished. Without empirical tests, a theory built only on a simulation model presents more a formal model of abstract phenomena than a theory of true social phenomenon. (p. 184)

Thus, the goal in the following sections is to provide an overview of two distinct cooperation and conflict models that have been produced from theory rather than specified directly from data. Beyond a mere review, I shall first provide an overview of each model's formalization to give insight into model construction using a theoretical (rather than data-driven) approach. I shall then outline each model's fundamental characteristics and emphasize how the model provides novel or deeper insight into how cooperation and conflict function as dynamic processes. Via example, the goal of this chapter is to provide insight to how dynamical systems models are useful in the broader research process. Before discussing the specific models, however, I will review some key concepts from dynamic systems theory relevant for the ideas that follow. Additional insights into dynamic systems concepts are provided in other chapters in this edited volume—see especially Peluso et al. (Chap. 3) and Yunzi-Tan, Strawinska-Zanko, and Michaels (Chap. 7).

6.1.2 Fundamental Dynamic Systems Concepts

Human behavior emerges from individual characteristics and situational factors that influence one another and behaviors, over time (cf. Funder 2006; Mishel 1990; Sherman et al. 2015). The person-situation interaction is fundamental to many contemporary social and personality psychological perspectives, and dynamic systems theory provides a coherent framework for better understanding the complex phenomena that are consequences of these evolving interactions. Cooperation and conflict are powerful examples of person-situation dynamics. For either of these interaction modes to emerge, there technically needs to be at least two actors involved in some sort of exchange (be it direct or indirect), and their exchange will be influenced by personal history, personality, affect, motivation, perceived similarities and differences, existing social bias (e.g., prejudice), as well as how the other actor responds in kind according to her or his constellation of individual qualities as they relate to the present situation, situational history, and future expectations. Clearly there is a lot at work in cooperation and conflict. How can dynamic systems perspectives help identify the most relevant factors at work in a given social exchange?

Essentially, a *dynamic system* is any set of interconnected elements that change over time (Miller and Page 2007; Vallacher et al. 2013). In social interactions, these elements can be any of the various person, situation, and behavioral components that are a part of the interaction. While the elements present in a social exchange may appear disorganized, a key property of dynamic systems is that they tend to self-organize. *Self-organization* is the tendency of a system (e.g., social interaction) to develop naturally emergent coherence as a product of the highly interactive factors that influence the system (Vallacher et al. 2013). As an example of these dynamic properties, deep friendships evolve as a unique type of cooperation typically involving reciprocity and self-disclosure (see Aron et al. 1997; Bukowski et al. 1994). The budding friendship involves exchange and discovery contingent on trust, which may relate to a person's sense of security (e.g., attachment style association with relationship quality, see Zayas and Shoda 2005). Depending on the friends' mutual behaviors exhibited as personal information is shared, the interaction can evolve into one of deeper intimacy and trust, thus exhibiting greater stability, or might evolve with surprise when irreconcilable incompatibilities or breaches of trust appear; this latter scenario might jeopardize the sustainability of the friendship. As many who have encountered a new situation with unfamiliar people know, social connectivity on the level of friendship cannot be forced—it emerges over time. The emergent nature of friendships reflects self-organization.

Over time, self-organization facilitates increased coherence between the elements in a system, which may be seen in the system exhibiting decreased variability. Self-organization leads to greater system stability through coherence (Guastello and Liebovitch 2009; Johnson 2001; Vallacher and Michaels 2012). When a system becomes progressively more stable over time, it is exhibiting attractor properties. An *attractor* is essentially a coherent state to which the given system returns, even after perturbation (Vallacher et al. 2013). Continuing with friendship as an example of cooperation, esteemed friends who have forged a steadfast bond exhibit an attractor state in their interaction by the friends having significant tolerance for one another when life difficulties emerge. The capacity for a true friendship to withstand the tests of life has been admired for centuries; indeed, a central theme in the Epic of Gilgamesh is that despite great differences at first, the protagonist forges a deep, enduring friendship in Enkidu who remains a beloved life companion. That friendships are so stable despite differences and difficulties alike reflects their attractor-like characteristics.

While the attractor-like dynamics found in friendships may be inspiring and valued in such relationships, attractors are not necessarily tied to solely harmonious forms of exchange. As dynamic systems concepts have been applied to cooperation and conflict theory and research, the attractor concept provides additional insight into the worst forms of human conflict (Coleman 2011; Vallacher et al. 2010a, b). In peace and conflict research, the worst conflicts are those that become stagnant with predictable, persistent patterns of overt hostility, incessant negative attitudes and emotions about opponents, and a sense that the resolution can never be obtained. These conflicts often become destructive and are said to be *intractable*; they are evident in the contemporary world with ongoing civil war in Syria, omnipresent tensions between Israel and Palestine, and unresolved disputes between people who share drastically polarized perspectives on issues like gun control and abortion (see Coleman 2011; Rouhana and Bar-Tel 1998). These and other intractable conflicts are sometimes said to be "frozen," as the destructive, malignant patterns of interaction between opposing sides are locked in place while also being highly resistant to change (e.g., Kriesberg 1980). Thus, a classic idea in cooperation and conflict theory and practice is that restoring flexibility (i.e., escaping the "conflict attractor") is a necessary antecedent to foster emergent peace (e.g., Bartoli et al. 2010).

While dynamic systems concepts, especially attractors, have been useful as theoretical tools for metaphorically understanding human interactions, there remains a need in the social sciences to go deeper and examine the dynamic properties of phenomena like cooperation and conflict, using more rigorous tools (Nowak et al. 1994; Vallacher et al. 2010b). Mathematical models afford one approach to operationalizing the ideas present in dynamic systems theories about human cooperation and conflict, which in-turn permits greater clarity in specifying problems and hypotheses worth pursuing via rigorous empirical studies. For many not formally trained in mathematics as a language and way of thinking, the translation of ideas and theories into elegant equations that capture key qualities of natural phenomena may seem mysterious. Thus, I shall apply an informal approach to examining two specific models relevant to cooperation and conflict in the remainder of this chapter. Drawing upon my own introduction to dynamic systems modeling, I will focus on using each model as an example of how qualitative concepts relevant in cooperation and conflict research and theory translate into the models without using traditional statistical approaches to model generation (e.g., regression, time-series analysis). Next, I will elaborate on each model's dynamical properties to deepen understanding of how dynamic systems metaphors relate to mathematical concepts. Finally, I will discuss how each model provides novel insights that complement existing knowledge and stimulate directions for future research.

6.2 Modeling Social Influence and Attitude Change

People rely on their attitudes for making decisions, forming inferences, and understanding themselves throughout much of their daily life. Attitudes hold such a powerful influence over people's behavior that they are frequently considered a fundamental factor in cooperation and conflict (Maio and Haddock 2007). From folk wisdom about solving a dispute by "settling differences" to theories and research across the social sciences, attitude similarity between people is generally considered a factor in harmonious social interactions (Deutsch 1973, 1994; Heider 1958; McNeel and Reid 1975), while attitude dissimilarity is considered a primary factor in conflict (e.g., Bodtker and Jameson 2001; Boulding 1961; Marcus 2006). However, in what way do shared similarities and differences contribute to cooperation or conflict in groups? Are certain types of attitudes more likely to generate the potential for conflict? Dynamical models have provided insightful answers to both of these questions.

Foremost, people use attitudes in everyday social exchange to express opinions and form understandings. When attitude similarity is noticed, people tend to feel more positively and are more likely to form stronger relations; in contrast, dissimilar attitudes tend to cause friction in the social exchange. Importantly, clashing attitudes are not always a source for problematic, destructive disputes. As Losada (1999) demonstrated with dynamical models based on the Lorenz equations, members in high-performance teams tend to exhibit a richer range of emotions and capacity for advocating their own perspectives (based on attitudes) while inquiring about other members' points of view. However, in some situations differently held attitudes factor into the development of a harmful conflict.

Dynamic systems ideas have been metaphorically applied to the development of destructive conflict. A key idea with this approach is how destructive, intractable conflicts arise from social interactions, behavior, and psychological processes taking on an attractor-like state (Coleman 2011; Vallacher et al. 2010a, b). As a conflict begins, the interacting adversaries have many degrees of freedom in terms of how they choose to interact, what behaviors they carry out, and whether they choose to give negative, aggressive emotions primacy or seek to find a way to calm the situation. As the dispute persists, however, degrees of freedom are lost, as negative attitudes about the situation become increasingly salient, expectations of conflict emerge, and a predictable, stable pattern of hostility and aggressive exchange takes root. In this scenario, the conflict emerges in the social system as an attractor. Once present, the conflict attractor becomes a powerful force, drawing adversaries into the dispute even if they try to cease hostilities. The conflict effectively becomes a way of life and becomes embedded in the social structure of those involved, locking the adversaries into a persistent pattern of continual dispute (Kriesberg 1980; Bartoli et al. 2010). Such destructive conflicts are evident across the world, including the tensions between Israel and Palestine (Coleman 2011; Kriesberg 1980), various enduring civil wars (e.g., Bartoli et al. 2010), and people who have strong attitudes about polarizing topics, like abortion and gun control (e.g., Coleman 2011). In order for conflict to assume this attractor-like state on a group level, members must share and organize around common attitudes. The members must adopt like-minded attitudes about a polarizing topic and about the opponents. In dynamic systems-inspired research, cellular automata have been applied to model the process by which people exchange attitudes, and this has provided insight to understanding how social influence plays a fundamental role in cooperation and conflict.

6.2.1 Background and Overview of Cellular Automata

In an intractable conflict, members of a particular group engaged in the conflict will typically share common attitudes about the opponent and situation; the conflict is embedded in public opinion. As Nowak and Lewenstein (1996b) explained, public opinion emerges from the interaction between factors on an individual and social level. Agent-based models have long been applied to studies of the interaction between individual (microlevel) and social (macro-level) phenomena. For example, von Neumann and Morgenstern's (1944) *Theory of Games and Economic Behavior* provided a foundation for studying which agent-based decisions were most likely to yield favorable outcomes for the individual in a structured, rule-based exchange (see Jones 2000; Monroe and Read 2008). Von Neumann and Morgenstern's ideas were influential in early simulation studies in economics, political science, and psychology, especially through the 1960s (Jones 2000), but have also held significance in understanding how cooperation and competition evolve. More recently, agent-based models similar to those used with early game theory research have provided significant insight into how cooperation is a more favorable general strategy of

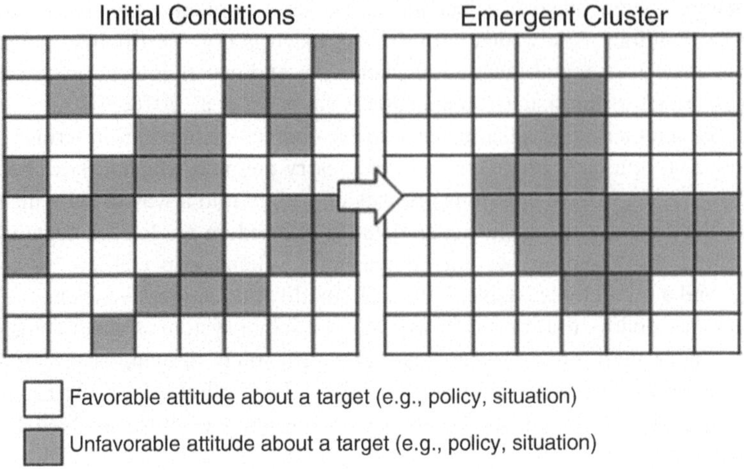

Fig. 6.1 A simple illustration of a cellular automata at the beginning of the simulation (left panel) and after some number of model iterations (right). Each cell in the 8 by 8 grid represents an agent. At the start of the simulation, left, the agents possess polarized attitudes and are randomly distributed in space. After some time, due to the model change rules, agents tend to organize based on similar attitudes, thus forming clusters (right)

interaction than competition, in social evolution (e.g., Axelrod 1984; Axelrod and Hamilton 1981; Nowak 2006). The key mathematical property of agent-based models is that they specify a set of mathematically expressed rules for how a collection of stationary "agents" (typically representing individual people) make decisions or transmit information from one agent to the next. Each agent is situated in a typically fixed topological space, which is usually an $n \times m$ grid, where each point in the grid represents an agent (see Fig. 6.1).

At the start of the model simulation, each agent is randomly assigned certain qualities relevant to the phenomenon of interest. For example, in social evolutionary models of cooperation and competition, some proportion of the agents are randomly assigned to be cooperators, whereas others are assigned to be defectors (i.e., agents who seek to selfishly compete with others). After the *initial conditions* that define the agents in their topological space are set, the simulation is run for some number of cycles (*iterations*). For each iteration, an agent may either retain its preexisting qualities or may be influenced by other, adjacent agents. For example, an agent may remain a cooperator or may switch interaction strategy to become a defector based on the influence of surrounding agents. Whether this switch occurs depends on the proportion of nearby agents who have similar or dissimilar qualities and on how the formulae governing the situation are specified. Agent-based models are thus intrinsically quite simple, rely on relatively elementary mathematical equations, but have proven powerful tools for understanding complex social phenomena and testing theories—especially as pertaining to large-scale social behaviors that otherwise might be difficult to examine scientifically, to include cooperation and conflict.

Cellular automata are one particular type of agent-based model that has been applied to examine attitude change and the propagation of conflict behaviors through a society. For example, Nowak et al. (1990) developed an early cellular automata model to capture properties of attitude change in a social setting. They developed the model according to principles outlined in Latané's (1981) social impact theory, a metatheory that characterizes social influence on a person's attitudes as a product of the number and proximity of other people, as well as how strongly the attitudes are held by one's self and by others. As Nowak et al. (1990) emphasized, translating social impact theory to a model requires that "all relations among variables must be exactly specified" (p. 366). Their modeling efforts began by considering a simple grid architecture as a topological space to simulate the interaction between agents. Noting that the influence between agents could be captured with a power function having a 0.5 exponent (per experimental studies), they next developed model equations:

$$i_s = N_s^{\frac{1}{2}} \Sigma \left[\frac{\left(\frac{s_i}{d_i^2} \right)}{N_s} \right] \tag{6.1}$$

The variables in the model include i_s as the overall influence an agent experiences from surrounding individuals, N_s representing the number of agents who share the same attitude as the target agent, s_i is how supportive the source of information is, and d_i as the distance between the source agent and recipient agent. Thus, this particular equation captures the influence of commonly shared, non-oppositional attitudes on an individual. A second equation of identical form was used to capture the influenced of opposing opinions on a single agent, except that in this case N_o represents the number of agents with the opposing attitudes and d_o reflects the strength of the source agent's persuasion. Note the change in the subscript from "i" to "o" merely keeps the individual equations distinct, where again one captures the influence of common shared attitudes among agents, while the other captures the influence of persuasion to change attitudes according to dissimilar others.

Although mathematically simple, this model captures core ideas from Latané's (1981) social impact theory. The key idea translated from the equation is that peoples' influence on a single person's attitudes is proportional to the number of people present, multiplied by the sum total of the number (variable N in the model) and strength of similar versus dissimilar attitudes (variable s in the model), divided by how far the different agents are from one another (variable d in the model). By dividing by the distance between people in the simulation, Nowak et al. (1990) were able to capture a key idea from social impact theory that the greater the distance between two people, the lesser their attitudes will impact one another (note that by dividing by distance, even fairly strong attitudes will be less influential when agents are quite distant from one another). Where is attitude change captured in this model?

Nowak et al. (1990) used a simple change rule to capture the phenomena; any given agent would match its attitudes to whichever group attitude was most strongly held among neighboring agents. Thus, if the attitudes held by those who are similar to the lone agent dominate, then the target agent maintains its attitudes. In contrast, if the surrounding group members exhibit stronger oppositional attitudes, then the target agent switches attitudes to match those of the neighbors.

6.2.2 Insights from Cellular Automata

Despite operating according to simple, arithmetically defined rules, cellular automata have proven insightful for understanding fundamental properties of human social behaviors, including cooperation and conflict. A key reason for this lies in the iterative nature of the models, meaning that they sequentially update agents' attitude states based on the previous model cycle (and, in the case of the first model iteration, based on the initial conditions). In their original article, Nowak et al. (1990) identified simple—but insightful—dynamics relating to the sustainment of majority versus minority opinions in a social group. They found that when a particular attitude was held by fewer than about 20 percent of group members, over time that particular attitude was almost always extinguished. Alternatively, polarized attitudes with the opposing vantages held by relatively equal proportions of group members (a 50/50 split) led to unpredictable dynamics, with one or the other attitudes gaining a majority dominance over time. Nowak et al.'s (1990) simulations revealed a novel finding: that minority attitudes held by 10–30 percent of members, spatially situated at the group's peripheries, tend to be resilient. While the minority attitudes became less present in the overall group, small sets of agents situated at the peripheries would form more stable bastions holding the minority attitude.

Following up on these preliminary results, Nowak and Lewenstein (1996a) further examined attitude change among interacting agents in a series of simulation studies. They delineated several findings from the model that translate to valuable insights for cooperation and conflict. Foremost, the simulations robustly revealed that of the course of interaction, the modeled agents did not tend to exhibit a random distribution with their held attitudes. Instead, agents tended to form "clusters," where agents having similar attitudes converged in the grid space. What this reveals is that in social situations, if minority opinions are held by a sufficient number of people, those opinions are unlikely to be eliminated. Instead, those of minority opinion are likely to group together, forming coalitions. This finding translates to cooperation dynamics when considered in conjunction with Nowak's (2006) studies on social evolution. Based upon Nowak's work, group selection processes favor those who are more likely to cooperate with others, in that like-minded individuals tend to form groups that over time become increasingly stable and resistant to change (see also Axelrod and Hamilton 1981; Vallacher and Brooks 2013). Through cellular automata, it is evident that such group-related phenomena are tied to attitude polarization.

Attitude polarization is one of the numerous factors that can transition a previously cooperative social interaction into one mired in conflict. While it is well-known that people who interact frequently tend to adopt similar attitudes over time, such as what occurs with the groupthink phenomena (e.g., Festinger et al. 1950; Janis 1982), Nowak and Lewenstein's (1996a) model added further depth to why the phenomena occur. In the simulations, agents who were more influential—holding stronger minority opinions—tended to become the nuclei for minority clusters. Once formed, these clusters tended to resist influence from the majority group, effectively operating according to a stable attractor state in which the minority opinion was a dominant force. The spatial distribution of the agents also mattered in the simulations. If a minority cluster was to remain stable, it needed a "leader," an agent that strongly held the minority attitudes. External influence from agents with opposing attitudes was likely to fail as the weaker minority cluster agents were able to reinforce their attitudes, thanks to the dominant leader. Nowak and Lewenstein (1996b) observed that in situations where several minority leaders emerge, the minority attitudes that are polarized against the larger majority tend to be particularly resilient, thus forming a "stronghold" for the minority attitudes. Finally, it matters how the leaders spatially relate to the minority agents, as in situations where the leaders are on the periphery of the minority attitude holding agents, the agents with the strongly held attitudes are able to "shield" the agents with weaker attitudes from the oppositional perspective. These model findings advance knowledge of cooperation and conflict as dynamic processes in multiple ways.

A key characteristic with intractable conflicts is that the disputes resist resolution, and when progress toward hostility cessation appears underway, the conflict often suddenly reemerges and quickly becomes reinstated. In these cases, the conflict is interlinked and embedded within the social context and may function as an attractor. Attempts to resolve the dispute fail, as the underlying polarizing norms held among combatants draw their ways of thinking and behaving back into the dispute (see Coleman 2011; Kriesberg 1980; Reiter 2009; Vallacher et al. 2010a, b). Cellular automata help to translate these somewhat abstract ideas into more tangible ways of understanding factors that may facilitate a sustained conflict, as powerfully demonstrated by a more recent model. Investigating how conflict develops and transitions into an intractable state, Nowak et al. (2010) applied a cellular automata model to Deutsch's (1973) Crude Law of Social Relations. As Nowak et al. (2010) explained, Deutsch's basic theoretical tenet was that cooperation leads to a constructive social interaction, whereas conflict leads to a destructive social interaction. However, these proposals were "crude" in the sense that they required the precision of mathematical formalization to better identify the process dynamics—to enable understanding of how cooperation and conflict lead to specific changes, what sorts of conditions lead to changes in cooperation and conflict, and identification of the situations in which the Crude Law is inapplicable. I suggest that Nowak et al.'s (2010) model elegantly formalizes Deutsch's ideas in a way that resolves many of these concerns, as may be observed by considering the fairly simple rules applied to the agents. In this model, competitive (conflict) behaviors promote formation of additional conflict that tends to spread in the simulation (society). However, following key ideas from evolutionary perspectives on cooperation and competition (e.g., Axelrod and Hamilton 1981; Caporael 2007; Nowak 2006), competitive behaviors tend to decay within the model over time. Importantly, Nowak et al. (2010) added parameters accounting for competitive interdependence between agents, which reflected whether one groups' interest impacted another's (in which case oppositional interests would be tied to conflict). In addition, psychological states were considered in terms of a general agent orientation toward cooperation or competition. Interestingly, no competitive orientations were included in the model's initial conditions, which again presumed an initially harmonious social system. From these simple rules, Nowak et al.'s (2010) simulation revealed that even isolated competitive interdependence and behavior can lead to the growth of an intractable conflict. While these parameters initially lead to little conflict, over time the absence of overt conflict masks a subversive formation of competitive psychological orientations among agents in vicinity of areas where the conflicting interests are present. This shift to negative, antagonistic orientations finally translates to the previously examined cellular automata forming, whereby agents possessing these competitive orientations cluster according to their polarized attitudes and in respect to the interdependence, and the conflict becomes a stable feature in the society for at least a given minority.

While there are many, deeper, insights from Nowak et al.'s (2010) research (see also Nowak et al. 2003), the key idea is that cellular automata capture critical features of how conflict emerges in society. In a society with predominantly positive, cooperative patterns of attitudes, behaviors, and exchange, a relatively small number of malcontent individuals who experience a competitive interdependence can become the source for spreading frustration and hostilities. As the situation goes unresolved, more individuals may be drawn into the discontent, leading to shifts in attitudes and other psychological processes, giving rise to increasingly destructive ways of thinking and behaving. Those who fall into such modes of thought and action self-organize and form clusters of like-mindedness, which can then emerge as an attractor phenomenon where the groups' oppositional perspectives and problematic behaviors become a fixture in the society. These model results fit well with the attractor metaphor for how destructive conflicts emerge from self-organizing inter- and intrapersonal processes in social situations (cf. Bartoli et al. 2010; Coleman 2011). More recently, Lim et al. (2007) used simulations to explore how the geospatial distribution of polarized groups contributes to social tensions and unstable intergroup relations that may facilitate conflict development. Their results were consistent with predictions afforded by cellular automata in that the formation of homogeneous clusters of like-minded agents in a sufficiently large area led to greater tendency for conflict emergence, which may be tied to a decrease in the complex connectivity among members of the distinct groups (e.g., Côté and Erickson 2004). Of interest, Lim et al. (2007) applied their model to the examination of violence in former Yugoslavia and in India, with their simulation successfully capturing the general pattern of conflict observed in these particular case studies.

In sum, cellular automata reveal that even complex social phenomena—such as cooperation and conflict—may emerge within a situation from relatively simple interaction rules that are sequentially iterated over time. This is a key feature of dynamic systems: the systems' behavior is inextricably tied to time, whereby past events exert complex and often nonlinear influence on the present observed situation. It is important to note that these events may result from external perturbations that impact the system or from factors internal to the people and their interactions. Cellular automata capture elements of these factors, considering agent distribution, attitude polarization, and strength of influence (or how strongly attitudes are held, see Latané and Nowak 1994). However, it is well-known that a multitude of interpersonal factors influence people's cooperative and competitive interactions, including communication styles (e.g., Gottman et al. 2002, Gottman and Peluso, Chap. 2), emotions (e.g., Nair 2007), and how individuals uniquely process information about social situations (see Pettit and Mize 2007, for review). Another type of model has proven useful for understanding how a combination of such factors combines with an interaction's time-dependent evolution, giving rise to various forms of cooperation and conflict.

6.3 The Nonlinear Emergence of Cooperation and Conflict

People rely on attitudes to understand their positions as they pertain to a myriad of issues relative to others, and from attitude similarity or difference, people are able to formulate judgments about social situations (see Kruglanski and Sleeth-Keppler 2007). While attitudes may be a fundamental psychological factor influencing cooperative versus competitive exchanges, a significant reason they prove influential over behavior is that they interface with other psychological processes, such as emotions. This interaction is evident, considering that attitudes may be defined as evaluations about a specific target and that attitudes consist of a combination of beliefs, emotions, and positive versus negative evaluations (see Maio and Haddock 2007). Emotions are thus an important component in attitude exchange and, hence, social interactions. However, there remains much to be understood about how emotions evolve and interact with various factors and contribute to cooperation and conflict as emergent social phenomena (cf. Kazan 1997; Nair 2007). For example, Coleman (2011) critiqued "standard approaches to research in the social sciences" for marginalizing research on the role emotions play in cooperation and conflict, noting "decades of research on social conflict has paid little attention to emotions" (p. 21). Thus, while cellular automata have proven insightful to how group-level dynamics unfold in a group-level situation where the interacting agents possess polarized attitudes, how do these polarized attitudes relate to affective dynamics between the interacting agents themselves? Addressing this question relates to the scale of the problem; cellular automata examine more macro-level processes between many individually modeled agents, allowing important insights into phenomena such as group organization and exchange contingent on differently held beliefs. In contrast, understanding the interface between attitudes and emotions requires examining behavior on a level that is more focused on microlevel factors, relating the factors in play between just two agents' interaction. Recognizing this, Liebovitch et al. (2008) applied a different type of model to better understand how cooperative, competitive, and mixed cooperative-competitive feedback evolve between what, from a cellular automata approach, would be considered only two interacting agents.

6.3.1 Background and Overview of the Nonlinear Two-Actor Model

The key idea behind Liebovitch et al.'s (2008) model is that any two agents (actors) experience an interaction that is based on some combination of factors relating to individual differences, conditions existing prior to the interaction onset, and whether the interaction is defined predominantly by cooperative, competitive, or mixed cooperative-competitive exchange. Before examining the model, its assumptions, and considerations in its development, it is worth noting that two-actor models have been developed to examine conflict scenarios. One good example is Richardson's (1935, 1951) model of large-scale war. As Hunt (2007) concisely described, Richardson reduced the numerous individuals and groups involved in large-scale conflict to just two interacting agents, each representing the two groups or nations engaged in war. The basic premise is that in a Cold War scenario, for instance, the interaction between two adversarial nations may worsen and become more plagued by conflict or improve and evolve toward peace based on three parameters: the amount of weaponry each nation possesses (x, y), how tolerant each nation is of the other's level of arms (X, Y), and how the nations react to one another (a, b). The equations are specified as differential equations; the notation to the left of the equal symbol effectively meaning x's state at any point in time is a result of (equals) the reaction (a) to the amount of arms y possesses relative to how much weaponry y is allowed to have (Y). Since the two nation-states are linked via their equations (x appears as a variable in the equation for nation y and y appears in the equation defining nation x), the equations are said to be coupled.

$$\begin{cases} \frac{dx}{dt} = a\left(y - Y\right) \\ \frac{dy}{dt} = b\left(x - X\right) \end{cases} \tag{6.2}$$

In this formulation, the presence of arms held by an adversary is threatening and may escalate tensions (thus the reaction is negative); however, the potential for such escalation is constrained by the extent to which some level of arms owned by an adversary are tolerated. Considering Gottman et al.'s (2002) models, this tolerance is similar to a damping parameter that helps constrain the potential for runaway conflict. What Richardson's model captured was the idea that in a Cold War scenario, the two interacting adversaries recognized that each perceived some level of nuclear arms was necessary to ensure self-protection and that the adversary was entitled to a similar level of arms. Such a scenario leads to a fragile status quo where tensions are kept in-check, and outright conflict is inhibited. However, the model also offered insight in that it demonstrated minor perturbations in the adversaries' level of arms (e.g., from perception of an unfair advantage, perhaps from the former Soviet Union having greater early success at developing long-range missiles) could trigger the opponent into ramping up levels of arms, leading to intensified conflict. In contrast, small-scale attempts to reduce arms would lead to the status quo destabilizing as the adversary responded with a similar gesture of disarming (see Hunt 2007, for a good summary). Richardson's model, though difficult to verify with quantitative data for various reasons, illustrated that parsimonious models capturing only the most critical features of social exchange between two actors may nevertheless prove insightful.

Parsimonious models are established from the logical construction of a specific theory that seeks a simple explanation of the phenomenon of interest based on consideration of only essential, fundamental factors (see Nowak 2004). In seeking a parsimonious model of cooperation and conflict, Liebovitch et al. (2008) developed equations that are similar to Richardson's (1935, 1951) formalization. In the two-actor conflict model, however, some important advancements appear. First, notice there appear two new constants, m and c. As Liebovitch et al. (2008, 2010) explained, m represents whether an actor is more or less prone to change relative to feedback from another person. Psychologically considered, this term makes sense in that some people experience stronger reaction to various phenomena relevant in a cooperation or conflict scenario, whereas others tend to exhibit modest response (e.g., emotional reaction, Larsen 2009). The c-term is a simple constant used to capture whether a person tends to feel more optimistic or pessimistic on accord of their own individual differences. Inclusion of these specific parameters reflects how this model was partially inspired by Gottman et al. (2002) (see Liebovitch et al. 2010, for discussion). The final components in the model are the parameters that capture how the two actors, x and y, are coupled—that is, how the two interact with one another over time. In the below equations, this coupling is defined by some function (a mathematical expression for how two variables relate), f. The notation simply indicates there is some mathematical way, f, to describe how actor y influences actor x and vice versa. The question here becomes exactly how do two people in a cooperative or competitive interaction influence each other? Are all levels of cooperation and competition equal? These challenging questions are critical for the development of a valid, workable model, and Liebovitch et al. (2008, 2010) demonstrated how importantly the careful consideration of precise, parsimonious assumptions about such phenomena contributes to the modeling process.

$$\begin{cases} \frac{dx}{dt} = m_1 x + b_1 + f_1(y,x) \\ \frac{dy}{dt} = m_2 y + b_2 + f_2(x,y) \end{cases} \tag{6.3}$$

The function that describes the interaction between two actors is critical for coupled nonlinear models, such as those described in this section, as the function will directly influence the outputs generated by the model. If not carefully specified, a function that is less well-chosen may yield results that are difficult to reconcile with real-world phenomenon. This is the case with Richardson's model (1935, 1951), as the function that describes how the two actors (*x*, *y*) interact is governed by linear terms. That is, in Richardson's model, one adversary's increasing arms will cause a direct, reciprocal increase in the other's arm stock—a proportional, tit-for-tat reaction. As evident from models applied to the evolution of cooperation, such model outputs can lead to simulation scenarios where competitive escalations are never ending (cf. Axelrod and Hamilton 1981; Nowak 2006). In considering how cooperative and competitive interactions evolve, Liebovitch et al. (2010) specified three qualitative considerations in the choice of model interaction functions. First, in scenarios involving relatively modest influence between the two actors, the researchers sought a function that captured people's tendency to engage in near equal levels of reciprocity. The presumption that people return small acts of kindness (e.g., holding a door open for someone) and respond mutually to minor slights (e.g., applying a car horn to scold an aggressive driver) fits with well-established principles about basic human behavior (e.g., Gouldner 1960; Perugini et al. 2003). A second property sought in the model interaction function was a naturally occurring limit. That is, as the interacting actors' exchange became increasingly cooperative or competitive, there should occur a point at which further escalation yields no appreciable change to the qualitative state of the social system. While not necessarily as intuitive, this second consideration makes sense when considering cooperation and competition extremes. For example, the worst escalation in a conflict might be the decimation of one's adversary; once an adversary is completely eradicated, would further escalation result in any change in the state of the conflict? As a final consideration, Liebovitch et al. (2008, 2010) sought a function that would permit representation of positive (cooperative) or negative (competitive) feedback; mathematically, having a function that permits both positive and negative parameter values would work with this consideration, noting many functions exist that will not operate with negative values. In conjunction with these logical considerations, ideas from research on aggressive behaviors provided further insight, with Pruitt (1969) and Bui-Wrzosinska (2005) identifying that conflict scenarios tend to evolve with fairly strong mutual feedback between adversaries at modest levels of dispute, but that as escalation takes place, there occurs a point at which no further aggression is realistically possible (e.g., no additional resources to devote to a war effort). Pruitt (1969) and Bui-Wrzosinska (2005) identified that these escalatory behaviors may be captured with sigmoid functions (Fig. 6.2), which commonly occur in natural systems (see Bejan and Lorente 2011).

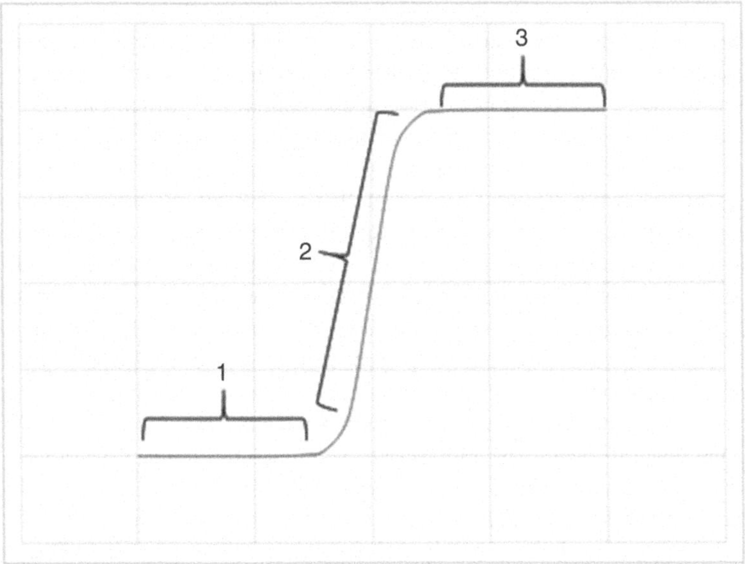

Fig. 6.2 Sigmoid functions capture many relationships in nature and tend to have two key properties useful in many modeling applications. First, they tend to allow the two interacting variables (x, y) to exert similar influence on one another when the interaction is defined with modest feedback (2). However, at higher levels of feedback within either variable, the influence reaches a limit where no further level of response will occur (1, 3)

Considering the logical, theoretical, and mathematical properties describing how escalating levels of cooperation or competition evoke reciprocal responses between two interacting people, groups, or nations, Liebovitch et al. (2008) specified that the coupling between the modeled actors should exhibit sigmoid function properties, captured by:

$$\begin{aligned} f_1(y,x) &= c_1 \tanh(y) \\ f_2(x,y) &= c_2 \tanh(x) \end{aligned} \tag{6.4}$$

The sigmoid function is defined by the hyperbolic tangent (*tanh*), where the coupling is thus defined—the state of actor x is dependent on whether the other person (y) is providing cooperative, positive feedback ($c_1 > 0$) or competitive, negative feedback ($c_1 < 0$). Effectively, the positive or negative feedback in this particular model is analogous to the type of reaction considered in Richardson's models (1935, 1951) (via the constants a and b). Combining the influence function with the remainder of Liebovitch et al.'s (2008) model, one obtains the complete simulation equations:

$$\begin{cases} \dfrac{dx}{dt} = m_1 x + b_1 + c_1 \tanh(y) \\ \dfrac{dy}{dt} = m_2 y + b_2 + c_2 \tanh(x) \end{cases} \tag{6.5}$$

As with cellular automata, this model operates by specifying some set of initial conditions that capture the state of the two actors' individual reaction tendency (m), general affective state (b), and type of interaction being shared with the other person (c). As with many models of behavioral phenomenon, it is difficult to discern precise values for these parameters, as the methods of measurement (e.g., Likert scales with self-report measures) may not translate easily to numeric data on a true ratio scale (which is what would be used in these types of mathematical models). However, the different constants may be specified with some reasonable validity by logically considering behavior from theoretical and empirical evidence. In their original simulations, Liebovitch et al. (2008) assigned a weakly negative value to capture people's reaction tendency, $m = -0.9$. This captures the idea that, in general, people are hesitant to instantaneously react to minor external stimuli that do not especially threaten existing ideas about goal pursuit, or the meaning of one's actions (e.g., Schwarz and Clore 2007; Vallacher and Michaels 2012; Vallacher and Wegner 2012), and general social expectations (e.g., tolerance for minor infractions considering the norm of reciprocity, see Axelrod and Hamilton 1981). A zero value was assigned for each actor's affective state ($b = 0$), representing that each individual begins the interaction in a relatively uninfluenced, neutral way of feeling, which may represent a person's subdued ambivalence about an unfamiliar social interaction that is not based on prior history between the people involved. Finally, the type of interaction shared between the two actors was altered across several different simulations to gain insight into what the simulation outputs captured in terms of different combinations of cooperation and conflict. Achieving this was possible by varying the intensity of the feedback ($-3 \leq c \leq 3$), where negative values represent an actor adopting a competitive interaction strategy, while positive values represent a cooperative strategy. Of interest, Liebovitch et al. (2008) explored the interaction not only in terms of different levels of mutual cooperation (or competition), but they examined what may happen in situations where one person strives to cooperate, while the other utilizes a competitive strategy. These mixed feedback scenarios led to particularly interesting dynamics, especially when considering the possibility that one actor changed whether cooperation or competition was employed as a strategy at some point during the interaction.

6.3.2 Insights from the Nonlinear Two-Actor Model

Several of the insights from the Liebovitch et al. (2008, 2010) model of two-actor cooperation-competition fit with what might be logically expected from human interaction. First, in cases where the two actors have only a weak exchange involving minimal feedback (i.e., the constant c is kept close to zero), the interaction evolves to a neutral state. As Liebovitch et al. (2010) further explained, the key to an interaction evolving to the neutral state is dependent on the relationship between the intensity of the feedback (c) and each actor's resistance to change (m). In any case where $c < m$, the interacting partners' hesitance to react will override other dynamics leading to minimal response by those involved in the exchange. These results seem externally valid considering that people pursue their own goals in daily life and will usually only deviate from those goals when there is sufficiently strong influence from other people, situational factors, or perhaps internal factors such as ego depletion (e.g., Bauer and Baumeister 2011). A second insight from the two-actor model that fits well with intuitive and empirical observations about human behavior is that once feedback becomes sufficiently strong ($|c| > |m|$), the final outcome of the interaction depends on the nature of the feedback. In a case where both actors exchange using cooperative, positive feedback, the duo evolves to a mutually successful (positive emotional) state or a mutually unsuccessful (negative emotional) state. This result makes sense, as it effectively captures a form of coordination, such as what occurs between people in a close relationship where mutual positive feedback (cooperation) leads to better outcomes, especially in times when problems are present, and mutual negative feedback (conflict) typically results in detrimental outcomes for both members of the couple (cf. Gottman 1993; Gottman et al. 2002). Finally, the third two-actor model result fitting well with general perspectives on social relationships is that the eventual outcome is contingent not only on the type of feedback utilized by the actors, but is linked to their individual initial state at the time the interaction begins. Liebovitch et al. (2010) framed this state in terms of emotions, such that given mutually shared positive feedback, two individuals who are initially in a positive affective state will eventually reach a positive end state; given an initial negative affective state, the two actors will evolve to a negative end state. In contrast, when the exchange is dominated by negative feedback, the actor who has the more positive affective state at the start of the interaction will tend to emerge a "winner," realizing a more positive outcome compared to the actor with the more negative affective state. Abundant evidence exists implicating that one's initial perceptions, which may be colored by emotions, influence behavior and subsequent outcomes. For example, Murray (1999) found that husbands' greater intrinsic negativity predicts a greater likelihood for divorce in married couples. On a more basic cognitive level, a negative, more pessimistic emotion at the start of a social exchange may alter perception and facilitate the well-known confirmation bias. Collectively, these three results seem to parallel general premises about interactions that are based on cooperation or conflict. However, the two-actor model's utility as a tool for exploring novel, perhaps unanticipated, outcomes is illustrated when considering the seldom explored possibility that in an interaction, one person may opt to cooperate, while the other competes. Moreover, the model is especially interesting when considering how people change their interaction styles during a social exchange.

In their simulations, Liebovitch et al. (2008, 2010) found that in a dyadic exchange, when one of the interaction partners adopts a cooperative (positive feedback) strategy and the other takes on a competitive (negative feedback) strategy, the two undergo a series of shifts in their affective state or in the state of their outcome over time. Effectively, each actor experiences periods of success, winning, and positive affect as well as periods of failure, loss, and negative affect. The oscillations between these two states eventually dampen to neutrality. These results are particularly interesting as they imply that in the case of conflict where both actors employ a competitive strategy, the actor who is beginning to lose the dispute may be able to alter the outcome by temporarily switching to a cooperative, positive feedback style. Strategically employing such a strategy for a brief period may allow the former loser to begin to gain an upper hand in the dispute, at which time resuming the competitive style might allow that actor to emerge dominant as the winner. Although tantalizing, these results do not entirely fit with what is presently known about patterns of cooperation and conflict, in part because scenarios, in which adversaries switch strategies, are not especially clear. Nevertheless, they do appear to match the general premise behind Osgood's (1962) graduated and reciprocated initiatives in tension reduction (GRIT) strategy, in which a protracted conflict might begin to be resolved by one of the two adversaries offering small concessions (weak positive feedback in the model's case) to alleviate the tensions. Over time, such concessions are thought to initiate increased cooperation via reciprocity norms that may lead to conflict resolution or at least emergence of a neutral truce. Indeed, in the Liebovitch et al. (2008, 2010) model, one actor's choice of adopting a cooperative strategy while the other seeks to compete results in the two reaching an eventual neutral end state, which may well be a better outcome for the person or group that emerges as the losing side in a conflict scenario.

These collective results may be visualized by considering phase space, which is basically a plot that reveals the state to which a given system will evolve over time considering many different initial conditions. A phase space may be generated for each of the scenarios Liebovitch et al. (2008, 2010) outlined and then interpreted graphically rather than strictly quantitatively. Although such an approach may run counter to analyses in a traditional social science sense, where inferential procedures and p-values are generally the norm, graphical analysis often proves useful in complexity science (see Hunt 2007; Liebovitch 1998). Two general phase space configurations from the two-actor model are presented in Fig. 6.3. The left panel, 3A, illustrates a scenario of weak feedback between the actors ($c < m$), whereby in time the two actors will reach a neutral form of interaction. This can be seen by following the trajectories (shown by the solid black lines) in the direction the lighter shaded arrows point. The interaction is neutral as all of the trajectories lead to the point at which $x = 0$ and $y = 0$, where zero indicates no significant affective response by either actor (actor x or actor y). This point of neutral interaction at the center of the graph is an attractor state. The right panel, 3B, shows a very different case. In this case, there is sufficiently strong positive feedback between the cooperating actors ($c > 0$ and $c > |m|$) leading both actors into a mutually positive state, which is an attractor (top right) or a mutually negative state (lower left).

All possible phase space configurations for the different types of feedback are presented in Fig. 6.4. Each panel from left to right shows a different phase space relative to whether actor x chooses to compete (left column of panels), remain neutral (middle column of panels), or opts to cooperate (right column of panels). From top to bottom are scenarios based on whether actor y elects to cooperate (top row), remain neutral (middle row), or compete (bottom row). Thus, at top left, the scenario portrays a situation in which actor y cooperates, while actor x competes.

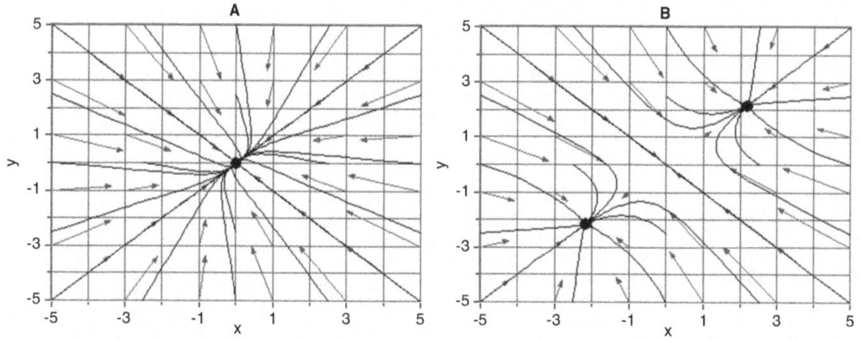

Fig. 6.3 Two phase space configurations from the two-actor model with attractors indicated by the black points. The solid, dark lines in each graph are trajectories that trace out how the actors' states (e.g., emotions) evolve over time in the interaction. The arrows indicate the direction of this evolution. Given sufficient time, the attractor in each scenario will be reached, at which point the actors' emotions will stabilize. Whether the end state is positive or negative depends on the values for each actor. For example, in 3B the bottom left attractor represents a mutually negative state as actor x has fairly negative state values here ($x \approx -1.5$), as is the same for actor y

By considering the different phase space configurations and the interesting results about interactions in which one actor opts for a cooperative strategy while the other utilizes a competitive strategy, it was apparent that the two-actor model could be experimentally tested using a simple procedure involving a controlled interaction between participants and trained research assistants who posed as participants (confederates). Thus, we carried out a study (Michaels et al. 2013) to examine what would happen in a scenario where people experienced a strictly cooperative, strictly competitive, or mixed cooperative-competitive pattern of feedback. In the experiment, participants ($N = 84$) initially completed a Likert scale survey to capture their agreement with 40 attitude statements, such as "nuclear weapons are necessary to protect our nation." The Likert scale could then be used to identify those items with which participants agreed and those with which they disagreed, allowing development of a script for a trained research assistant to use in a future controlled social exchange. For the exchange situation, participants returned to the lab and took turns reading, and listening to, the same 40 statements with a research assistant posing as an actual participant. However, instead of merely answering the survey a second time, participants and the assistant expressed whether or not they each agreed with the statement. In doing this, we could generate a sense of cooperation (both agreed and disagreed with the same items) or competition (one person agreed with the statement, while the other disagreed). Finally, we could generate mixed feedback scenarios by having the confederate switch midway through the interaction from sharing similar attitudes with the participants to having dissimilar attitudes (or vice versa). By measuring how much participants' attitudes changed between the pre-interaction, to the period during the interaction, and by measuring their emotional states immediately after the interaction session, we obtained results that largely fit with model predictions. The greatest exception was in the mixed

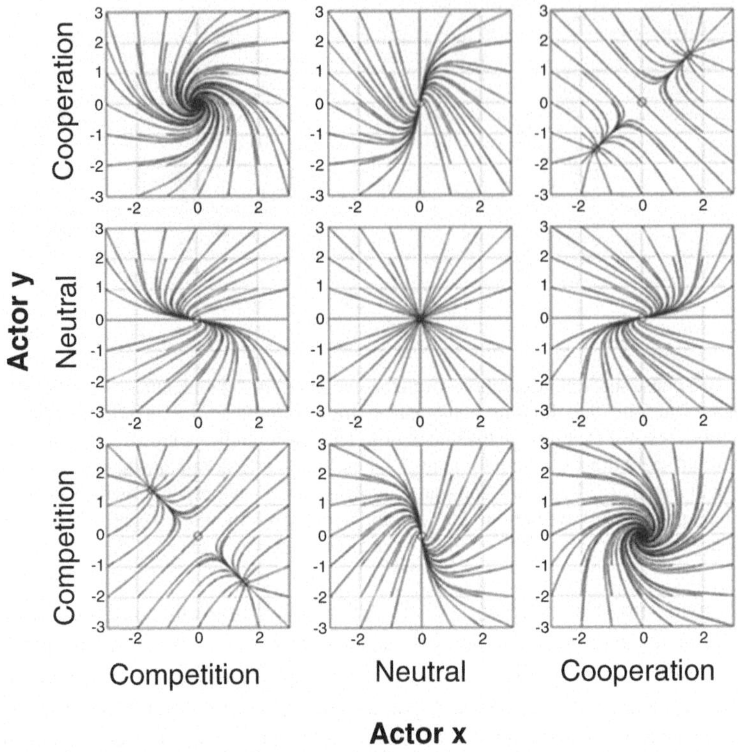

Fig. 6.4 Main phase space configurations from the Liebovitch et al. (2008, 2010) two-actor model of cooperation-competition. Columns show outcomes relative to actor *x*'s interaction strategy, while rows depict outcomes relative to actor *y*'s

feedback cases, where simulations suggested the onset of one actor engaging in positive feedback while the other engaged in negative feedback should have resulted in their experiencing state oscillations; that is, mixed feedback should have caused perturbations in attitudes and emotions (representing relevant psychological states). The expected perturbations only emerged in scenarios where the confederates began the interaction session with cooperative, positive feedback and then shifted to a competitive, negative feedback in the middle of the exchange. In this situation, participants reacted strongly, exhibiting strong attitude change and markedly more negative emotions. In contrast, a scenario where the confederate transitioned from a competitive to a cooperative feedback style, little attitude or affective response occurred within the participants. Thus, the interesting model finding—that the onset of mixed feedback in an already competitive exchange could perhaps perturb people toward an eventually more neutral outcome or perhaps reverse who is the winner versus loser in a conflict—may not be fully accurate with actual interactions. These findings about mixed feedback scenarios in the experiment fit with the general notion that the transition from an established cooperative exchange to one that is competitive may be particularly unsettling to people. For one, this sort of interaction transition would likely be experienced negatively (as our affect data validate), and negatively perceived experiences tend to be more salient and influential over people's thoughts and actions (e.g., Taylor 1991, cf. Fredrickson 2013). In addition, the strong cognitive and affective response to sudden negative feedback may reflect a mode of automatic thought, such as the likelihood that the emergent negative feedback triggers an affective alert circuit in the brain, causing participants to shift their attention to a possible problem as a means of adaptation to the evolving situation (cf. Isbell et al. 2013; Schwarz and Clore 2007).

The two-actor model combined with our experimental results provides a glimpse into important dynamics that emerge during the course of a social interaction and may help to advance ideas in conflict resolution. Metaphorically, intractable conflicts appear to operate as effective attractor states, where people's thoughts, feelings, and actions become frozen in a persistent destructive pattern that ultimately resists attempts at resolution (Coleman 2011; Vallacher et al. 2010a, b). If the solution to such disputes is to "defreeze" the system and work to get those involved to develop more flexible ways of thinking, feeling, and acting (Bartoli et al. 2010; Coleman 2011), then it is critical to see evidence of such flexibility in a situation where conflict is replaced with cooperative strategies. While the two-actor model predicts this flexibility will emerge when a lone actor ceases competition and adopts cooperation, the experiment results suggest otherwise. Per the experiment, once competition is present in an exchange, developed agreement offered on the part of just one individual (in this case the confederate) does little to alter participants' existing attitudes or emotions. Meanwhile, once competition takes root in a previously cooperative exchange, participants react swiftly with attitude perturbation and subsequent negative emotions. These findings warrant further study, especially in light of conflict resolution practice. For example, what would happen if a progressively increased mode of cooperation replaces competition? Is there a certain point at which sufficiently strong cooperation, offered unilaterally, might tip the interaction out of the conflict? Such questions might offer new insights into Osgood's (1962) GRIT strategy. Another unexplored domain within this context is whether the dynamics change once a third party is introduced to a situation based in conflict, which no doubt adds increased complexity to the situation (Goldstein and Freeman 1990). Considering how negotiations are common to conflict resolution, this domain of research would be especially welcome. Beyond traditional negotiation strategies, it would be especially interesting to explore whether qualities intrinsic to the third party play a role in conflict alleviation; for instance, there is evidence that some particularly destructive disputes transform thanks to the unique qualities of social entrepreneurs—individuals who introduce generally informal, small-scale changes in a given social environment that help to resolve endemic social problems, like conflict (e.g., Bartoli et al. 2010; Praszkier and Nowak, 2012). As is true of any productive line of scholarship, combining mathematical models with empirical research leads to exciting insight and new questions. This ebb and flow of discovery and inquiry promises to advance our understanding of cooperation and conflict as dynamic processes, and this iterative progression is only enhanced by carefully crafted models.

6.4 Models as an Integrative Tool in Research

Cooperation and conflict are natural outcomes in societies populated by people who have individual differences, unique motives and backgrounds, and seek to achieve their goals in situations with inequalities and scarce resources. Both phenomena are inexorable components in the story of human civilization, perhaps tied to fundamental social evolution (Axelrod and Hamilton 1981; Nowak 2006). Cooperation is a foundation for social stability, and while some degree of conflict may be helpful for stimulating social change and evoking action, at a certain threshold it takes on a dark and destructive characteristic. The powerful sway that cooperation and conflict hold over human lives is reflected in these two forces prominently appearing in literature, art, and forms of expression that originate with the earliest known human civilizations. Despite cooperation and conflict having a certain centrality in life, how the two evolve over time and transition from one to the other remains poorly understood. Fortunately, new ways of thinking about cooperation and conflict are taking root in interdisciplinary social sciences and hold promise of advancing new discoveries and integrating fragmented, disciplinary isolated empirical and applied research.

Dynamic systems-inspired approaches to theory, research, and practice combined with mathematical models offer tools that may help better advance deep understanding of cooperation and conflict as processes. As Kenrick et al. (2009) concluded, "dynamical systems research holds the promise of understanding how the thoughts and motivations of individuals add up to monumental group level processes from civil rights marches to international conflicts" (p. 497). Trying to see such macro-level processes as they relate to the microlevel elements intrinsic within any single individual is quite difficult and is akin to "seeing the forest for the trees." However, as illustrated with the two examples in this chapter, mathematical models are often helpful for understanding phenomena that may, on the surface, appear too complex to grasp from a rigorously scientific perspective. Cellular automata models help to reveal how basic principles of interaction between individual people (agents) translate to large-scale group organization. Such organization helps to understand how phenomena such as polarization and conflict might emerge in societies. Nonlinear differential equation-based models are useful for understanding how different variables intrinsic to individuals combine with different forms of social interaction to give rise to cooperation or conflict over time. Importantly, neither of these examples is superior to the other or to other approaches used in mathematical modeling. Instead, each modeling method brings with it unique strengths and weaknesses, affording insights into different domains of a complex world.

Mathematical models are further useful for interdisciplinary research in the sciences, where different disciplines and subfields share complex and dynamic relations in their own right. Mathematical models serve a central role in scientific inquiry as they are derived on parsimony, forcing researchers to specify exacting properties of the phenomenon of interest (Hunt 2007; Liebovitch et al. 2010; Nowak 2004). Furthermore, mathematics is a common language enabling models, developed upon premises unique to certain disciplines, to foster interdisciplinary collaborations. For example, a sociologist or political scientist may not be familiar with the different regulatory focus styles (Higgins 1998) and how a person's orientation toward promoting gains in a social exchange versus preventing losses might alter the emergence of cooperation or conflict, but translating this idea with simple equations could permit the basic premise behind the relationships to be shared with anyone who is willing to examine the mathematics. The capacity to generate interdisciplinary teams to work at difficult questions with multiple perspectives, distinct forms of knowledge, and discipline-specific tools is doubtlessly reshaping science. As numerous scholars indicate, interdisciplinary research is an emergent norm in the sciences (e.g., Porter and Rafois 2009; Sloman 2008; Van Leeuwen and Tijssen 2000), and the social sciences may be well positioned as a particular hub for interdisciplinary connectivity (Boyack et al. 2005). Mathematical models are essential tools for capitalizing on interdisciplinary team-based research.

Of course, no technique is perfect. As Richardson (1960) admonished, models may generate spurious outputs that are illogical; the simplification of complex phenomena necessary for producing many types of models may be over-reductive, eliminating the qualitative patterns of greatest interest from inclusion. As Nowak et al. (1994) further articulated, measurement in social and behavioral research is inherently imprecise, rendering a differential equation-based modeling approach "too ambitious for social psychology" (p. 285). Given such limitations of measurement and "weaknesses of those tools" (Nowak et al. 1994, p. 280) used in quantitative modeling, qualitative models may better serve research in the social and behavioral sciences. Thus, as with any resources in a well-equipped scientific toolbox, models are but one device that may be used to enhance understanding and foster insights into promising new directions for theory and research. Part of the utility models afford for the scientific process derives from the fact that through precise specification of the factors relevant in cooperation and conflict, well-designed models naturally permit operationalization of variables and provide further clarity through the causal relationship between variables. As illustrated with the two-actor model, this clarity can translate to robust experimental design. Thus, models help to translate ideas from theories, which may be especially important since the success of dynamic systems approaches in the social sciences requires advancing beyond the use of key concepts, like attractors, as mere metaphorical tools (Nowak and Vallacher 1998).

In many respects, great works of literature endure through time as they carry metaphorical messages relevant to anyone's life. The timeless nature of cooperation and friendship, of conflict and great battles, reemerge in countless stories, epics, and legends. It is only in relatively recent years, however, that tools afforded to social scientists are allowing increased precision in studying these complex phenomena. Models are just one of these great tools. Returning to the Epic of Gilgamesh, themes in this ancient epic can be seen in the two models discussed at the core of this chapter. If the characters in this story were transformed into agents in a cellular automata model, there is no doubt Enkidu, described as a "wild man," would have held a minority attitude among more "civilized" peers. It is thus unsurprising that, isolated from like-minded others, Enkidu eventually found himself unable to return to his old ways and thus joined Gilgamesh and his people. Observing the two-actor model—and upon first meeting Gilgamesh and Enkidu engaged in a great battle—the duo certainly began with negative, competitive-oriented feedback, striving to win. However, when it was evident Gilgamesh had won the battle, Enkidu surrendered and acknowledged his opponent's victory. In this way, Enkidu initiated a change in the style of feedback, embracing a positive, cooperative feedback style, perturbing the interaction out of the conflict and initiating a path toward friendship. If the ideas from these two models may be applied to interpret one of the oldest written works known in civilization, what insights may they—and yet to be formulated models—provide for understanding current and future conflicts? It is my hope that the ideas in this chapter might inspire an enterprising social scientist to explore mathematical modeling and perhaps reach out to colleagues in other disciplines, mathematics included, to embark on new directions of scientific discovery. Certainly there is much that remains to be written about the power and promise of mathematical models in understanding cooperation and conflict.

References

Aron, A., Melinat, E., Aron, E. N., Vallone, R. D., & Bator, R. J. (1997). The experimental generation of interpersonal closeness: A procedure and some preliminary findings. *Personality and Social Psychology Bulletin, 23*, 363–377. https://doi.org/10.1177/0146167297234003

Axelrod, R. (1984). *The evolution of cooperation*. New York: Basic Books.

Axelrod, R., & Hamilton, W. D. (1981). The evolution of cooperation. *Science, 211*, 1390–1396. https://doi.org/10.1126/science.7466396

Bartoli, A., Bui-Wrzosinska, L., & Nowak, A. (2010). Peace is in movement: A dynamical systems perspective on the emergence of peace in Mozambique. *Peace and Conflict: The Journal of Peace Psychology, 16*, 211–230. https://doi.org/10.1080/10781911003691609

Bauer, I. M., & Baumeister, R. F. (2011). Self-regulatory strength. In K. D. Vohs & R. F. Baumeister (Eds.), *Handbook of self-regulation: Research, theory, and applications* (2nd ed., pp. 64–82). New York: The Guilford Press.

Bejan, A., & Lorente, S. (2011). The constructal law origin of the logistics S curve. *Journal of Applied Physics, 110*, 024901. https://doi.org/10.1063/1.3606555

Bodtker, A. M., & Jameson, J. K. (2001). Emotion in conflict formation and its transformation: Application to organizational conflict management. *The International Journal for Conflict Management, 12*, 259–275. https://doi.org/10.1108/eb022858

Boulding, K. E. (1961). *Conflict and defense*. New York: Harper & Brothers.

Boyack, K. W., Klavans, R., & Borner, K. (2005). Mapping the backbone of science. *Scientometrics, 64*, 351–374. https://doi.org/10.1007/s11192-005-0255-6

Bui-Wrzosinska, L. (2005). The dynamics of conflict in a school setting. Unpublished master's thesis. Poland: University of Social Sciences and Humanities.

Bukowski, W. M., Hoza, B., & Boivin, M. (1994). Measuring friendship quality during pre- and early-adolescence: The development and psychometric properties of the friendship qualities scale. *Journal of Personal and Personal Relationships, 11*, 471–484. https://doi.org/10.1177/0265407594113011

Burns, D. (2007). *Systemic action research: A strategy for whole system change*. Bristol: The Policy Press.

Caporael, L. R. (2007). Evolutionary theory for social and cultural psychology. In A. W. Kruglanski & E. T. Higgins (Eds.), *Social psychology: Handbook of basic principles* (2nd ed., pp. 3–18). New York: The Guilford Press.

Chatrchyan, S., (CMS Collaboration), et al. (2013). Study of the mass and spin-parity of the Higgs Boson candidate via its decays to Z Boson pairs. *Physical Review Letters, 110*, 081803–1–081803-15. https://doi.org/10.1103/PhysRevLett.110.081803

Coleman, P. T. (2011). *The five percent: Finding solutions to seemingly impossible conflicts*. New York: Public Affairs.

Côté, R. R., & Erickson, B. H. (2004). Untangling the roots of tolerance: How forms of social capital shape attitudes toward ethnic minorities and immigrants. *American Behavioral Scientist, 47*, 740–765. https://doi.org/10.1177/0002764209331532

Deutsch, M. (1973). *The resolution of conflict: Constructive and destructive processes*. New Haven: Yale University Press.

Deutsch, M. (1994). Constructive conflict resolution: Principles, training, and research. *Journal of Social Issues, 50*(1), 13–32. https://doi.org/10.1111/j.1540-4560.1994.tb02395.x

Deutsch, M., Coleman, P. T., & Marcus, E. C. (2006). *The handbook of conflict resolution: Theory and practice*. San Francisco: Wiley.

Festinger, L., Schachter, S., & Back, K. (1950). *Social pressures in informal groups*. Stanford: Stanford University Press.

Fredrickson, B. L. (2013). Updated thinking on positivity ratios. *American Psychologist, 68*, 814–822. https://doi.org/10.1037/a0033584

Funder, D. C. (2006). Towards a resolution of the personality triad: Persons, situations, and behaviors. *Journal of Research in Personality, 40*, 21–34. https://doi.org/10.1016/j.jrp.2005.08.003

Gleick, K. (1987). *Chaos: Making a new science*. New York: Penguin.

Goldstein, J. S., & Freeman, J. R. (1990). *Three-way street: Strategic reciprocity in world politics*. Chicago: University of Chicago Press.

Gottman, J. M. (1993). The roles of conflict engagement, escalation, and avoidance in marital interaction: A longitudinal review of five types of couples. *Journal of Consulting and Clinical Psychology, 61*, 6–15. https://doi.org/10.1037/0022-006X.61.1.6

Gottman, J. M., Murray, J. D., Swanson, C. C., Tyson, R., & Swanson, K. (2002). *The mathematics of marriage: Dynamic nonlinear models*. Cambridge, MA: MIT Press.

Gouldner, A. W. (1960). The norm of reciprocity: A preliminary statement. *American Sociological Review, 25*, 161–178. https://doi.org/10.2307/2092623

Guastello, S. J., & Liebovitch, L. S. (2009). Introduction to nonlinear dynamics and complexity. In S. J. Guastello, M. Koopmans, & D. Pincus (Eds.), *Chaos and complexity in psychology: Theory of nonlinear dynamics* (pp. 1–40). New York: Cambridge University Press.

Guastello, S. J., Koopmans, M., & Pincus, D. (Eds.). (2009). *Chaos and complexity in psychology: Theory of nonlinear dynamics*. New York: Cambridge University Press.

Gunion, J. F., Haber, H. E., Kane, G., & Dawson, S. (1990). *The Higgs hunters' guide*. Cambridge, MA: Perseus Publishing.

Heider, F. (1958). *The psychology of interpersonal relations*. New York: Wiley.

Heiman, G. W. (2001). *Understanding research methods and statistics: An integrated introduction for psychology* (2nd ed.). Boston: Houghton Mifflin Company.

Higgins, E. T. (1998). Promotion and prevention: Regulatory focus as a motivational principle. *Advances in Experimental Social Psychology, 30*, 1–46. https://doi.org/10.1016/S0065-2601(08)60381-0

Hunt, E. (2007). *The mathematics of behavior*. New York: Cambridge University Press.

Inzlicht, M., & Schmeichel, B. J. (2012). What is ego depletion? Toward a mechanistic revision of the resource model of self-control. *Perspectives on Psychological Science, 7*(5), 450–463. https://doi.org/10.1177/1745691612454134

Isbell, L. M., Lair, E. C., & Rovenpor, D. R. (2013). Affect-as-information about processing styles: A cognitive malleability approach. *Social and Personality Psychology Compass, 7*(2), 93–114. https://doi.org/10.1111/spc3.12010

Janis, I. L. (1982). *Groupthink: Psychological studies of policy decisions and fiascos* (2nd ed.). Boston: Houghton-Mifflin.
Johnson, S. (2001). *Emergence: The connected lives of ants, brains, cities, and software*. New York: Scribner.
Jones, A. (2000). *Game theory: Mathematical models of conflict*. Cambridge, UK: Woodhead Publishing.
Kenrick, D. T., Neuberg, S. L., & Cialdini, R. B. (2009). *Social psychology: Goals in interaction* (5th ed.). New York: Pearson.
Kazan, M. K. (1997). Culture and conflict management: A theoretical framework. *The International Journal of Conflict Management, 8*, 338–360. https://doi.org/10.1108/eb022801
Kriesberg, L. (1980). Interlocking conflicts in the Middle East. *Research in Social Movements, Conflicts and Change, 3*, 99–118.
Kruglanski, A. W., & Sleeth-Keppler, D. (2007). The principles of social judgment. In A. W. Kruglanski & E. T. Higgins (Eds.), *Social psychology: Handbook of basic principles* (2nd ed., pp. 116–137). New York: The Guilford Press.
Larsen, R. J. (2009). Affect intensity. In M. R. Leary & R. H. Hoyle (Eds.), *Handbook of individual differences in social behaviors* (pp. 241–254). New York: The Guilford Press.
Latané, B. (1981). The psychology of social impact. *American Psychologist, 36*, 343–356. https://doi.org/10.1037/0003-066X.36.4.343
Latané, B., & Nowak, A. (1994). Attitudes as catastrophes: From dimensions to categories with increasing information. In R. R. Vallacher & A. Nowak (Eds.), *Dynamical systems in social psychology* (pp. 219–250). San Diego: Academic Press.
Liebovitch, L. S. (1998). *Fractals and chaos: Simplified for the life sciences*. New York: Oxford University Press.
Liebovitch, L. S., Naudot, V., Vallacher, R. R., Nowak, A., Bui-Wrzosinska, L., & Coleman, P. T. (2008). The dynamics of two-actor cooperationcompetition conflict models. *Physica A: Statistical Mechanisms and Its Applications, 387*, 6360–6378. https://doi.org/10.1016/j.physa.2008.07.020
Liebovitch, L. S., Vallacher, R. R., & Michaels, J. (2010). Dynamics of cooperation-competition interaction models. *Peace and Conflict: Journal of Peace Psychology, 16*, 175–188. https://doi.org/10.1080/10781911003691625
Lim, M., Metzler, R., & Bar-Yam, R. (2007). Global pattern formation and ethnic/cultural violence. *Science, 317*, 1540–1544. https://doi.org/10.1126/science.1142734
Losada, M. (1999). The complex dynamics of high performance teams. *Mathematical and Computer Modelling, 30*(9–10), 179–192. https://doi.org/10.1016/S0895-7177(99)00189-2
Maio, G. R., & Haddock, G. (2007). Attitude change. In A. W. Kruglanski & E. T. Higgins (Eds.), *Social psychology: Handbook of basic principles* (pp. 565–586). New York: Guilford.
Marcus, E. C. (2006). Change and conflict: Motivation, resistance, and commitment. In M. Deutsch, P. T. Coleman, & E. C. Marcus (Eds.), *The handbook of conflict resolution: Theory and practice* (pp. 436–454). Hoboken: Wiley.
McNeel, S. P., & Reid, E. C. (1975). Attitude similarity, social goals, and cooperation. *Journal of Conflict Resolution, 19*, 665–681. https://doi.org/10.1177/002200277501900409
Michaels, J. L., Vallacher, R. R., & Liebovitch, L. S. (2013). Volatile psychological dynamics in social interactions: Attitudes and emotions react asymmetrically to interaction shifts between agreement and disagreement. *Social Psychological and Personality Science, 4*, 705–713. https://doi.org/10.1177/1948550613482985
Miller, J. H., & Page, S. E. (2007). *Complex adaptive systems: An introduction to computational models of social life*. Princeton: Princeton University Press.
Mishel, W. (1990). Personality dispositions revisited and revised: A view after three decades. In L. Pervin (Ed.), *Handbook of personality: Theory and research* (pp. 111–134). New York: Guilford Press.
Monroe, B. M., & Read, S. J. (2008). A general connectionist model of attitude structure and change: The ACS (attitudes as constraint satisfaction) model. *Psychological Review, 115*, 733–759. https://doi.org/10.1037/0033-295X.115.3.733

Murray, S. L. (1999). The quest for conviction: Motivated cognition in romantic relationships. *Psychological Inquiry, 10*(1), 23–34. https://doi.org/10.1207/s15327965pli1001_3

Nair, N. (2007). Towards understanding the role of emotions in conflict: A review and future directions. *International Journal of Conflict Management, 19*, 359–381. https://doi.org/10.1108/10444060810909301

Nowak, A. (2004). Dynamical minimalism: Why less is more in psychology. *Personality and Social Psychology Review, 8*(2), 183–192. https://doi.org/10.1207/s15327957pspr0802_12

Nowak, M. A. (2006). Five rules for the evolution of cooperation. *Science, 314*, 1560–1563. https://doi.org/10.1126/science.1133755

Nowak, A., & Lewenstein, M. (1996a). Modeling social change with cellular automata. In R. Hegselmann, U. Mueller, & K. Troitszch (Eds.), *Modeling and simulation in the social sciences from the philosophy of science point of view*. New York: Springer.

Nowak, A., & Lewenstein, M. (1996b). Modeling social change with cellular automata. In R. Hegselmann (Ed.), *Modeling and simulation in the social sciences from a philosophical point of view* (pp. 249–285). Boston: Kluwer.

Nowak, A., & Vallacher, R. R. (1998). *Dynamical social psychology*. New York: Guilford.

Nowak, A., Szamrej, J., & Latané. (1990). From private attitude to public opinion: A dynamic theory of social impact. *Psychological Review, 97*, 362–376. https://doi.org/10.1037/0033-295X.97.3.362

Nowak, A., Vallacher, R. R., & Lewenstein, M. (1994). Toward a dynamical social psychology. In R. R. Vallacher & A. Nowak (Eds.), *Dynamical systems in social psychology* (pp. 279–293). San Diego: Academic Press.

Nowak, A., Vallacher, R. R., & Miller, M. E. (2003). Social influence and group dynamics. In T. Millon & M. J. Lerner (Eds.), *Handbook of psychology, volume 5: Personality and social psychology*. New York: Wiley.

Nowak, A., Deutsch, M., Bartkowski, W., & Solomon, S. (2010). From crude law to civil relations: The dynamics and potential resolution of intractable conflict. *Peace and Conflict: Journal of Peace Psychology, 16*, 189–210. https://doi.org/10.2139/ssrn.1493525

Osgood, C. E. (1962). *An alternative to war or surrender*. Urbana, IL: University of Illinois Press.

Perugini, M., Gallucci, M., Presaghi, F., & Ercolani, A. P. (2003). The personal norm of reciprocity. *European Journal of Personality, 17*, 251–283. https://doi.org/10.1002/per.474

Pettit, G. S., & Mize, J. (2007). Social-cognitive processes in the development of antisocial and violent behavior. In D. J. Flannery, A. T. Vazsonyi, & I. D. Waldman (Eds.), *The Cambridge handbook of violent behavior and aggression* (pp. 322–343). New York: Cambridge University Press.

Porter, A., & Rafois, I. (2009). Is science becoming more interdisciplinary? Measuring and mapping six research fields over time. *Scientometerics, 81*, 719–745. https://doi.org/10.1007/s11192-008-2197-2

Praszkier, R., & Nowak, A. (2012). *Social entrepreneurship: Theory and practice*. Cambridge: Cambridge University Press.

Pruitt, D. G. (1969). Stability and sudden change in interpersonal and international affairs. *Journal of Conflict Resolution, 13*, 18–38. https://doi.org/10.1177/002200276901300102

Pruitt, D. G., & Olczak, P. V. (1995). Beyond hope: Approaches to resolving seemingly intractable conflict. In B. B. Bunker, J. Z. Rubin, and associates (Eds.), *Conflict, cooperation, and justice: Essays inspired by the work of Morton Deutsch* (pp. 59–89). San Francisco: Josey-Bass Publishing.

Reiter, Y. (2009). *National Minority, regional majority: Palestinian Arabs versus Jews in Israel*. Syracuse: Syracuse University Press.

Richardson, L. F. (1935). Mathematical psychology of war. *Nature, 135*, 830–831. https://doi.org/10.1038/1361025a0

Richardson, L. F. (1951). Could an arms-race end without fighting? *Nature, 168*, 567–568. https://doi.org/10.1038/168567b0

Richardson, L. F. (1960). *Arms and insecurity*. Chicago: Quadrangle.

Rouhana, N., & Bar-Tel, D. (1998). Psychological dynamics of intractable conflicts: The Israeli-Palestinian case. *American Psychologist, 53*, 761–770. https://doi.org/10.1037/0003-066X.53.7.761

Schwarz, N., & Clore, G. L. (2007). Feelings and phenomenal experiences. In A. W. Kruglanski & E. T. Higgins (Eds.), *Social psychology: Handbook of basic principles* (2nd ed., pp. 385–407). New York: The Guilford Press.

Sherman, R. A., Rauthmann, J. F., Brown, N. A., Serfass, D. S., & Jones, A. B. (2015). The independent effects of personality and situations on real-time expressions of behavior and emotion. *Journal of Personality and Social Psychology, 109*(5), 872–888. https://doi.org/10.1037/pspp0000036

Sloman, A. (2008). Putting the pieces together again. In R. Sun (Ed.), *The Cambridge handbook of computational psychology* (pp. 684–709). New York: Cambridge University Press.

Strogatz, S. H. (1994). *Nonlinear dynamics and chaos: With applications to physics, biology, chemistry, and engineering*. Cambridge, MA: Perseus Books Publishing.

Sun, R. (Ed.). (2008). *The Cambridge handbook of computational psychology*. New York: Cambridge University Press.

Taylor, S. E. (1991). Asymmetric effects of positive and negative events. The mobilization-minimization hypothesis. *Psychological Bulletin, 110*, 67–85. https://doi.org/10.1037/0033-2909.110.1.67

Vallacher, R. R., & Brooks, C. (2013). Adaptation and coherence: Evolutionary and dynamical perspectives on human violence. In T. K. Shackelford & R. D. Hansen (Eds.), *The evolution of violence* (pp. 187–209). New York: Springer.

Vallacher, R. R., & Michaels, J. L. (2012). Why the mind works: The emergence of consciousness from mental dynamics. In S. Kreitler & O. Maimon (Eds.), *Consciousness: Its nature and functions*. Hauppauge: Nova Science Publishers.

Vallacher, R. R., & Wegner, D. M. (2012). Action identification theory. In P. Van Lange, A. W. Kruglanski, & E. T. Higgins (Eds.), *Handbook of theories in social psychology* (pp. 327–348). Los Angeles: Sage.

Vallacher, R. R., Coleman, P. T., Nowak, A., & Bui-Wrzosinska, L. (2010a). Dynamical foundations of intractable conflict: Introduction to the special issue. *Peace and Conflict: The Journal of Peace Psychology, 16*, 113–125. https://doi.org/10.1080/10781911003691294

Vallacher, R. R., Coleman, P. T., Nowak, A., & Bui-Wrzosinska, L. (2010b). Rethinking intractable conflict: The perspective of dynamical systems. *American Psychologist, 65*, 262–278. https://doi.org/10.1037/a0019290

Vallacher, R. R., Michaels, J. L., Wiese, S., Strawinska, U., & Nowak, A. (2013). Mental dynamism and its constraints: Finding patterns in the stream of consciousness. In D. Cervone, M. Fajkowska, M. W. Eysenck, & T. Maruszewski (Eds.), *Personality dynamics: Meaning construction, the social world, and the embodied mind*. New York: Eliot Werner Publications.

Vallacher, R. R., Van Geert, P., & Nowak, A. (2015). The intrinsic dynamics of psychological process. *Current Directions in Psychological Science, 24*(1), 58–64. https://doi.org/10.1177/0963721414551571

Van Leeuwen, T., & Tijssen, R. (2000). Interdisciplinary dynamics of modern science: Analysis of cross-disciplinary citation flows. *Research Evaluation, 9*, 184–187. https://doi.org/10.3152/147154400781777241

von Neumann, J., & Morgenstern, O. (1944). *Theory of games and economic behavior*. Princeton: Princeton University Press.

Westaby, J. D., Pfaff, D. L., & Redding, N. (2014). Psychology and social networks: A dynamic network theory perspective. *American Psychologist, 69*, 260–284. https://doi.org/10.1037/a0036106

Zayas, V., & Shoda, Y. (2005). Do automatic reactions elicited by thoughts of romantic partner, mother, and self relate to adult romantic attachment? *Personality and Social Psychology Bulletin, 31*, 1011–1025. https://doi.org/10.1177/0146167204274100

Chapter 7
A Dynamical Approach to Conflict Management in Teams

(Rae) Yunzi Tan, Jay L. Michaels, and Urszula Strawinska-Zanko

7.1 A Dynamical Approach to Conflict Management in Teams

Conflict is a natural and inevitable aspect of teams in organizational life. As individuals organize into collective units (i.e., teams) to coordinate tasks, develop innovative products, or solve complex problems, it is common for them to experience conflict with one another in terms of disagreements, opposing viewpoints, and incompatible activities (Jehn 1994, 1995, 1997; Wall and Callister 1995). Such prevalent experiences of conflict in teams can also be both beneficial and detrimental to the teams' overall functioning and effectiveness (for reviews and meta-analyses, see De Dreu and Weingart 2003; de Wit et al. 2012; Jehn and Bendersky 2003; O'Neill et al. 2013).

Indeed, past research has found that under certain conditions and depending on the type of conflict involved (i.e., task, relationship, and process conflict), conflicts in teams, also referred to as intragroup conflicts, can broaden members' perspectives and promote critical thinking (Amason et al. 1995; Nemeth 1995) and, in turn, enhance team decision quality (e.g., Schulz-Hardt et al. 2002) or team innovation

(Rae) Y. Tan (✉)
School of Public and International Affairs, College of Public Affairs,
University of Baltimore, Baltimore, MD, USA
e-mail: rtan@ubalt.edu

J. L. Michaels
Department of Psychology, University of South Florida Sarasota-Manatee, Sarasota, USA

U. Strawinska-Zanko
Department of Conflict Resolution Studies, College of Arts, Humanities, and Social Sciences,
Nova Southeastern University, Fort Lauderdale, FL, USA

U. Strawinska-Zanko, L. S. Liebovitch (eds.), *Mathematical Modeling of Social Relationships*, Computational Social Sciences,
https://doi.org/10.1007/978-3-319-76765-9_7

(e.g., De Dreu 2006; De Dreu and West 2001). At the same time, team conflicts are also likely to lower team trust (e.g., Jehn et al. 2008; Polzer et al. 2002; Rispens et al. 2007), dampen member satisfaction (e.g., Amason and Schweiger 1997), and undermine team performance (e.g., Carnevale and Probst 1998; Staw et al. 1981). In order to reconcile these mixed findings, scholars have begun to shift toward examining the role of internal group processes and mechanisms, such as *team conflict management*, in promoting the benefits of intragroup conflict, while mitigating its deleterious effects (e.g., Behfar et al. 2008; DeChurch and Marks 2001; DeChurch et al. 2013).

In this chapter, we focus on the importance of team conflict management as an influential means through which conflict can be effectively addressed or handled in teams. Here, we define team conflict management as the approaches, behaviors, and activities that members of a team employ to deal with conflict that has occurred or can occur within the team (Deutsch 1973; Alper et al. 2000; DeChurch and Marks 2001). Namely, members in teams may address or handle internal conflicts in the following ways: seek mutual gains and uncover shared interests among one another (i.e., cooperation); compel others to yield to their demands or accept their positions (i.e., competition); and ignore the tensions or withdraw from dealing with disagreements (i.e., avoidance; e.g., Chen et al. 2005).

Undoubtedly, the concept of team conflict management is not new; an abundant body of research knowledge has been accumulated over the past several decades to inform our understanding about the nature, antecedents, and consequences of managing conflict in teams (e.g., Alper et al. 2000; Behfar et al. 2016; Jones and White 1985; Weider-Hatfield and Hatfield 1995). Yet, as research and scholarship on team conflict management grew over the years, substantial gaps and limitations remain in the literature.

More specifically, our current knowledge of team conflict management has been largely represented by static, fixed views of the phenomenon, due to the widespread adoption of the positivist-reductionist paradigm and the linear methodological and statistical tools typically employed in the social sciences. With some exceptions (see Behfar et al. 2008, 2016; Kuhn and Poole 2000; Poole and Dobosh 2010), empirical studies on team conflict management have primarily relied on using classic and field experimental designs (e.g., DeChurch et al. 2007; Jordan and Troth 2004; Kleinman et al. 2003; Paul et al. 2005) and, to a greater extent, on cross-sectional survey designs (e.g., Alper et al. 2000; Chen and Tjosvold 2002; DeChurch and Marks 2001; O'Neill and Allen 2014; Tjosvold et al. 2006; 2009); consequently, these research strategies have created a linear, piecemeal picture of what, in reality, is a dynamic, multifaceted, and adaptive process.

Fortunately, recent advances in mathematics and computing, along with a paradigmatic shift toward dynamism in the social sciences (Guastello et al. 2009; Nowak and Vallacher 1998; Vallacher and Nowak 2007; Vallacher et al. 2010), have presented novel and more sophisticated tools for us to examine team conflict management in a gestalt and dynamic manner. Over the past decade, we have begun to see more research in related domains pay greater attention to the multilevel and temporal aspects of these constructs, such as group conflict asymmetry (e.g., Avgar and Neuman 2015; Jehn et al. 2010), intractable social conflict dynamics (e.g., Coleman et al. 2007; Vallacher et al. 2010), and team emergent states (e.g., Marks et al. 2001; Okhuysen and Richardson 2007). Social science researchers are also steadily and increasingly adopting the tools and methodologies associated with complexity thinking (Holland 1995; Strogatz 2003; Weisbuch 1992) and dynamical systems theory (Nowak and Vallacher 1998; Vallacher et al. 2002a) to help us more fully capture and understand the inherent complexities and dynamics associated with social-psychological phenomena.

Building upon these recent trends, we reason that the time is now ripe for us to integrate the existing fragmented landscape of team conflict management research into a more coherent whole and to extend our current knowledge by taking into account the temporal and dynamic aspects of managing conflict in teams. The main goal of this chapter, then, is to offer some promising insights into the utility and relevance of applying the dynamical systems perspective in our study of team conflict management. In this chapter, we provide arguments for why team conflict management should be treated as a dynamical system and how key properties and characteristics associated with dynamical systems can be observed in how team members manage conflicts. We also present preliminary findings in our research on team conflict management dynamics to provide some initial observations and tentative conclusions about the complex, adaptive, and dynamical nature of team conflict management and its association with team performance and effectiveness. Relatedly, our research seeks to address three key questions: (a) Do different conflict management approaches (e.g., cooperation, competition, and avoidance) employed by individual team members combine to create meaningful, discernable patterns for the team as a whole over time and in ways that differentiate high-performing teams from low-performing ones? (b) Can key characteristics associated with dynamical systems, such as emergence and attractor states, be observed in a team conflict management system? (c) How do external events or conditions shape the internal dynamics and eventual impact of a team conflict management system?

This chapter is organized into four sections. The first section briefly reviews the current research literature on team conflict management and outlines some of the key gaps that we aim to address in our research. The second section provides an expanded account of our application of the dynamical systems framework to our study of team conflict management, with a focus on how key dynamical system properties and characteristics, such as emergence and attractor states, are likely to occur in a dynamical team conflict management system. In the third section, we describe our current and ongoing research efforts to test our propositions and ideas associated with dynamical team conflict management and report our preliminary findings associated with this work to date. Finally, we summarize our tentative conclusions and their implications for theory, research, and practice, and conclude with some recommendations for future research.

7.2 Current Research on Team Conflict Management

For more than three decades, we have seen a steady accumulation of research and knowledge on how conflict is managed in organizational work teams (e.g., Alper et al. 2000; Behfar et al. 2008; DeChurch et al. 2013). Much of this literature has also built upon existing theory and research on conflict management at the interpersonal level, that is, how one handles disagreements, tensions, or incompatibilities with another party (e.g., Alper et al. 2000; DeChurch and Marks 2001; Farmer and Roth 1998; Jarboe and Witteman 1996; Somech 2008). For instance, past studies on conflict management in work groups and teams have typically relied on classic conflict management theories and typologies, such as Deutsch's (1949, 1973) theory of cooperation and competition, Rahim and Bonoma's (1979) model of conflict handling styles, Thomas and Kilmann's (1974) dual-concern model, and Van de Vliert and Euwema's (1994) meta-taxonomy of conflict management, to conceptualize conflict management among members at the group level. Among these theories and typologies, Deutsch's theory of cooperation and competition has been the most commonly used conceptual framework in a majority of the scholarly literature on team conflict management, as evidenced in Tjosvold and colleagues' extensive body of research on the topic (e.g., Alper et al. 2000; Tjosvold 1990, 2006, 2008; Tjosvold et al. 2009, 2014). For this reason, we focus on Deutsch's theory in this chapter and its application in past research, as well as expound on its relevance to our treatment of team conflict management in dynamical terms.

In Deutsch's (1973) theoretical discussion of cooperative and competitive processes in conflict situations, he contends, "the crux of the differences between cooperation and competition lies in the nature of the way the goals of the participants in each of the situations are linked (p. 20)." Put otherwise, how the goals between two parties involved in a conflict situation, whether they are perceived or real, tend to shape the parties' approaches toward resolution or handling of the conflict. When conflict parties view their goals as cooperatively linked, they are more likely to communicate with each other openly and honestly, share information that accentuates their similarities and common interests, adopt more friendly and trusting attitudes toward each other, and frame their conflicting interests as a mutual problem to be solved (Deutsch 1973, 2014). In contrast, parties who view their goals as competitively linked are likely to limit information exchange, adopt more suspicious and hostile attitudes toward each other, experience heightened sensitivities to differences between them, and adopt more coercive strategies to "defeat" the other in order to accomplish one's interests (Deutsch 1973, 2014).

Since its inception, Deutsch's (1949, 1973) theory of cooperation and competition has been extensively examined and tested in numerous empirical studies, and its theoretical tenets further extended by other scholars, such as Johnson and Johnson (2005, 2011) and Tjosvold (1984, 1998). Specifically, existing studies have found that cooperative approaches to conflict in teams tend to be associated with higher levels of affective trust (Hempel et al. 2008), conflict efficacy (Alper et al. 2000), effective decision-making (Kuhn and Poole 2000), organizational justice

(Chen and Tjosvold 2002), team innovativeness (Tjosvold et al. 2009), and team performance (Tjosvold et al. 2006), when compared to relatively competitive approaches. The latter mode of conflict management, however, could also yield positive organizational outcomes in some instances (e.g., Liu et al. 2009).

In the organizational domain, past research on cooperation and competition has also identified the occurrence and role of conflict avoidance, another general approach to managing conflict, which has been typically linked to competitive and independent goals among conflict parties (Tjosvold and Sun 2002). Unlike cooperative and competitive approaches, however, current findings on the effects of avoidant approaches to conflict have been more limited and much less clear. While a number of studies have pointed to non- or negative associations between conflict avoidance and work outcomes (e.g., Chen and Tjosvold 2002; Tjosvold et al. 2009), few others have also noted positive effects (e.g., Bear et al. 2014; De Dreu and Van Vianen 2001; Kuhn and Poole 2000; Tjosvold and Sun 2002).

Taken altogether, the current research literature has found that cooperative approaches to conflict, relatively to competitive and avoidant approaches, are more likely to yield both short- and long-term benefits for groups and teams in organizations across a variety of situations (e.g., Chen and Tjosvold 2002; Chen et al. 2005; Kuhn and Poole 2000; O'Neill et al. 2013; Tjosvold et al. 2009). And it is only under certain circumstances that competitive and avoidant approaches are likely to lead to favorable team outcomes (e.g., De Dreu and Van Vianen 2001; Tjosvold and Sun 2002).

Based on past work, we now know aplenty about various antecedents and consequences of cooperative and competitive conflict management (to a lesser extent for conflict avoidance) in teams. Yet, the current research literature continues to reflect disparate, disjointed pieces in the team conflict management puzzle. One notable piece missing in this puzzle has to do with the relative dearth of research that examines the dynamic, adaptive, and complex aspects of the phenomenon. For instance, past research has generally conceptualized teams as using cooperative, competitive, or avoidant approaches in conflict situations (e.g., Alper et al. 2000; Tjosvold et al. 2009), and this implicitly assumes that members within teams share similar conflict management approaches or that they converge to similar approaches to address internal team conflicts. This assumption is also reflected in the ways the construct has been operationalized, i.e., using group means scores that are derived from the aggregation of team members' scores on an instrument assessing team conflict management (e.g., Boros et al. 2010; Chen and Tjosvold 2002; De Dreu and Van Vianen 2001).

Researchers have also generally relied on using cross-sectional designs in both experimental (e.g., DeChurch et al. 2007; Jordan and Troth 2004; Jones and White 1985; Oetzel 1999) and survey studies (e.g., Alper et al. 2000; Boros et al. 2010; DeChurch and Marks 2001; Desivilya and Yagil 2005), which have culminated in data gathered at a single point in time during the teams' life spans. Such data necessarily constrain our view of the dynamism and complexity inherent in the process of managing conflict among team members and have consequently led to a somewhat limited understanding of the phenomenon at large. More recently, however, scholars have started to shift toward using more exploratory and emergent designs (e.g., Behfar et al. 2008, 2016; Poole and Dobosh 2010) to better capture the nuances and dynamism of managing conflict in teams. While this research trend is certainly encouraging, more needs to be done to accelerate our efforts in pushing the field along this direction. To that end, we discuss how the dynamical systems framework may be instrumental in furthering our research knowledge of team conflict management, particularly in relation to its adaptive and dynamic nature over time, in the next section. Specifically, we outline relevant fundamental properties of dynamical systems and argue how team conflict management, when viewed as a dynamical system, is likely to exhibit these properties. We believe that by examining team conflict management through the dynamical systems lens, we will be able to build a more holistic understanding of how this phenomenon occurs, changes, and evolves over time to affect critical outcomes at various levels of analysis.

7.3 Toward a Dynamical Perspective of Team Conflict Management

In this section of the chapter, our hope is to inspire the reader to see team conflict management as a process evolving over time that can be fruitfully investigated from the perspective of dynamical systems. As evidenced in several other chapters of this volume, this perspective has been successfully used to investigate complex phenomena at different levels of personal and social reality ranging from intra-psychological processes, such as self-esteem and self-regulation (Nowak and Vallacher 1998), to more complex social dynamics, such as protracted conflict (Vallacher et al. 2013). Our conceptualization of team conflict management as a dynamical process employs several concepts from dynamical systems approach that characterize all living systems. These key concepts include time evolution, self-organization, and attractor dynamics, and they will be reviewed in the upcoming sections to provide a necessary context for the subsequent presentation of the application of dynamical systems to research on team conflict management.

7.3.1 Time Evolution

To conceptualize a social or psychological phenomenon as a dynamical system means to define it as a set of interconnected elements (e.g., beliefs, feelings, behaviors, people) that influence each other and change over time. Changes in individual elements underlie the evolution of the entire system and identification of meaningful patterns at the system level becomes a fascinating and challenging task for anybody trying to understand and perhaps predict the behavior of the system. "If a reliable pattern can, in fact, be discerned, the unit of analysis is no longer the discrete changes constituting the pattern, but rather the pattern itself" (cf. Nowak and Vallacher 1998, p. 33). An illustrative example might be the dynamics of opinions in social groups where beliefs held by individual people impact the attitudes of others through social influence processes and the lower-level interactions between people lead to the emergence of such configurations at the group level as attitude polarization or clustering. In case of team conflict management, we look at the behaviors, attitudes, and feelings of individual team members and seek discernible configurations at the team level that emerge when team members work together toward achieving their common goal through fruitful collaboration, resolution of competition among their ideas, adaptive avoidance of disputes, as well as a wide range of other more or less productive responses to the demands of the current situation they face as individuals and as a team. Our goal is to shed some light on the role of temporal influences in team conflict management approaches.

7.3.2 Self-Organization

Undoubtedly, every team is unique in a number of respects (e.g., individual members' characteristics, history); however, as a team, the team members function as interdependent elements of a larger system. Their interdependence stems from the fact that realization of the interests of one individual has implications for the interests of another individual and the entire group (Nowak et al. 2010). Individuals within the team influence each other's thoughts, behaviors, and feelings and develop consistent ways of relating to one another. We look at those mutual influences in terms of self-organization processes that underlie the emergence of order in a system with many interacting elements. Thus, through self-organization, complex behaviors, such as team performance, can emerge simply through the interaction of the individual group members in the context of their inter- and intrapersonal processes; no formal external pressures are necessary for the team dynamics to develop (e.g., Losada 1999; Nowak et al. 2003). In the course of interactions, individuals adjust to one another, and highly coherent patterns of thought, affect, and behavior may emerge. These highly coherent structures function as the system's attractors (Vallacher et al. 2013).

7.3.3 Attractor Dynamics

An attractor is a system's preferred state; in other words, it is a state that the system will frequently return to over time or after change. In psychological and social systems, attractor-like behavior can be observed in many contexts, for example, habit formation, automatic forms of thinking, or human preference for predictability and routine. For the purpose of this chapter, we briefly discuss the basic properties of a system governed by attractor dynamics and explain how they can be recognized in empirical data.

There are many different forms of attractors; the one most commonly found and perhaps easiest to recognize in social psychological systems is the fixed-point attractor. In a social psychological system, a fixed-point attractor describes a tendency of individual's thoughts, feelings, or behaviors to return to a certain restricted range of possible options. The key feature of the dynamics observed in systems that are governed by an attractor is their tendency to stabilize and return to the preferred state after perturbation, which does not mean that such systems are stuck in an invariable, rigid steady state. Rather, attractor dynamics mean that the final behavioral response of the system is shaped by the unique interaction of external factors (e.g., new information) and the intrinsic attractor forces. For example, research has shown that individuals with high self-esteem tend to focus on positive self-relevant information and discard negative feedback, which suggests that positively evaluated self-relevant thoughts function as an attractor for the stream of thoughts of persons with high self-esteem. At the same time, impactful external events, such as a major failure or achievement, may impact their self-esteem level by lowering or increasing it, but these effects will be transient, and after some time, the level of self-esteem will return to the range associated with the existing attractor (Vallacher et al. 2002a). Similar fixed-point attractor dynamics has been found in empirical studies of mood regulation (e.g., Larsen and Diener 1985), depression (e.g., Johnson and Nowak 2002), or interpersonal conflict (e.g., Coleman 2011; Michaels et al. 2013a).

In experimental data, the existence of a fixed-point attractor in a system is signaled by the convergence of relevant variables on particular values over time. A system may also have more than one fixed-point attractor in which case the system is described as bistable (with two attractors) or multi-stable (more than two attractors). The presence of multiple attractors will be manifested in empirical data by a more complex pattern of a temporal trajectory of the variables in question usually characterized by oscillations between two (or more) preferred states. In an analysis of our empirical data, we utilize the definitional characteristics of a fixed-point attractor to discern such preferred, i.e., often visited, states and related internally generated dynamics. If our empirical records show that the variable in question, i.e., conflict management approach, no longer changes but becomes stable over time, this implicitly suggests the presence of an attractor. This can happen if, for example, team members with initially diverse conflict management strategies adopt the same conflict management approach, e.g., cooperation, in the course of their interactions and mutual influences. A system may also achieve stability by adopting a more complex configuration with, for example, certain members of the team, over time, becoming competitive and other team members adapting by becoming cooperative and/or avoidant. Attractor-like processes have been identified in various psychological and social systems but, to the best of our knowledge, have never been explored in the context of team conflict management research in the way proposed by our team.

7.3.4 *Intrinsic Dynamics and External Influences*

One of the reasons the dynamical perspective is so well suited to the analysis of team conflict management or any interpersonal processes, for that matter, is its emphasis on recognizing and understanding the time-dependent relation between external influences and a system's internally generated forces (Vallacher et al. 2002b). Therefore, in addition to understanding what patterns of conflict management approaches emerge when individuals work as a team, in our research, we also strive to better understand how such emergent patterns relate to team functioning. To do that, we correlate the observed patterns of internal team dynamics (e.g., convergence of individual conflict management approaches on cooperation) with subjective and objective measures of team performance. Our goal is to determine if distinct patterns of conflict management approaches are linked to team performance, i.e., whether the low-performing teams differ from the high-performing teams in the dynamical patterns of conflict management they develop over time. Further, we are interested to see whether these patterns, should they emerge, vary in relation to external influences or critical junctures throughout the teams' life spans.

To sum, in our research approach, we recognize teams as dynamical systems comprised of interacting and interdependent individuals, and we want to understand what distinct configurations of conflict management approaches are formed during the team life span, how such configurations change over time, and what implications team conflict management dynamics have for team functioning. In order to do that, we collect rich data regarding team members' conflict management approaches over an extended period of time. We next employ novel tools to analyze the temporal trajectories of changes for each of the key variables and to identify meaningful patterns. Emphasis on the temporal patterns rather than final outcomes is a distinguishing feature of dynamical systems approach, and in the context of team conflict management research, our exploration has a potential to enrich the existing body of knowledge by providing a detailed account of the dynamics of managing conflict in teams. The study presented in the next section of the chapter combined conventional experimental methods and statistics with novel dynamical systems-inspired techniques to investigate how team members' conflict management styles evolve during the course of their interactions leading to a completion of a team project.

7.4 Method

7.4.1 *Participants*

For our preliminary research, we recruited voluntary participants from a graduate course at a public university located in the southeastern USA. A total of 12 students, who were enrolled in the course and randomly organized into 3 project teams (4 students per team), participated. Each team worked on the same project assignment over the course of 8 weeks in the semester. The project assignment required the teams to conduct an organizational assessment and to provide recommendations based on their assessment findings. Ten (83.3%) of the participants were female, and 11 (91.7%) indicated the USA as their country of origin. On average, the participants' ages ranged from 25 to 34 years old, and they had 10 to 14 years of work experience. Two (16.7%) of the participants identified themselves as White/Caucasian, eight (66.7%) identified as Black/African-American, and the remaining two (16.7%) identified as other (e.g., biracial, multiracial). All participants provided informed consent prior to participating in the study.

7.4.2 Design and Procedures

In this research, we used a longitudinal, repeated-measures survey design to collect the data. Participants were asked to complete a series of survey questionnaires during class time at seven consecutive time intervals. Specifically, participants completed a survey measure assessing their conflict management approaches in their teams for each of the first six time intervals; on the seventh and last time interval, they completed a survey measure assessing their perceptions of team effectiveness. A member of the research team administered the surveys in paper format to the participants at the end of each class period over 7 weeks. Participants completed the paper surveys and then returned them to the researcher before leaving the classroom. It took approximately 20–30 min for each participant to complete the surveys each week.

7.4.3 Variables and Measures

Conflict management approaches To assess the conflict management approaches used by individual members within the project teams, we adapted and used an existing scale that measured cooperative, competitive, and avoidant conflict management (Chen et al. 2005; Tjosvold et al. 2009). Participants were asked to rate all 12 items in this measure on a five-point Likert scale (1 = "strongly agree"; 5 = "strongly disagree"). Sample items in this measure include: "I work with my teammates to find a solution that will be good for all of us," "I try to persuade my teammates to make concessions without necessarily making concessions myself," and "I try to keep differences of opinion quiet." The Cronbach alphas, averaged across all six time intervals, for the overall scale and for the three subscales assessing cooperative, competitive, and avoidant conflict management dimensions, respectively, are reported as follows: $\alpha_{overall} = 0.66$, $\alpha_{cooperative} = 0.54$, $\alpha_{competitive} = 0.62$, and $\alpha_{avoidant} = 0.71$.

Team effectiveness In this study, we assessed team effectiveness using both a self-report and an objective measure. For the self-report measure, we used the

Comprehensive Assessment of Team Member Effectiveness (CATME) scale developed by (Ohland et al. 2012). Participants were asked to rate their own and other team members' level of effectiveness on a list of 33 behaviors and attitudes using a five-point Likert scale ("1 = strongly disagree"; "5 = strongly agree"). Sample items in this measure include: "Did a fair share of the team's work"; "Exchanged information with teammates in a timely manner"; "Helped the team to plan and organize its work"; "Cared that the team produced high-quality work"; and "Had the skills and abilities that were necessary to do a good job." The Cronbach alpha for this team effectiveness scale was 0.97. For the objective measure of team effectiveness, we obtained both the students' individual course grades and the final group project grades from the course instructor at the end of the semester.

Participant demographics At Time 7 (or Week 7), we also asked participants to provide information about their age, gender, racio-ethnicity, country of origin, as well as years of former and current work experience.

7.5 Results

Our primary goal with the present study was to apply dynamical systems theory and techniques to better understand the dynamic interplay between interpersonal processes and external events that are associated with team outcomes. While our small sample size, 12 participants, does not permit robust conclusions, we intend the present research to be an exploratory study that applies established inferential statistical procedures alongside dynamical systems-inspired techniques to contribute new insights in how team performance is fundamentally tied to social, psychological, and situational events that interact over time. Accordingly, our exploratory work is relevant as it serves as an illustration of how extending traditional cause-effect research paradigms into designs that better acknowledge the importance of time can generate new insight into complex group phenomena. We thus sought to address three primary questions with our study. Foremost, do individual team members' cooperative, competitive, and avoidant approaches to conflict combine to create meaningful, discernible patterns at the team level that distinguish more versus less successful teams? Second, can we observe attractor states in the team dynamics, and if so, are such states associated with team outcomes? Finally, how do external events perturb the team members and alter their individual approaches to conflict?

An initial examination of team members' cooperative, competitive, and avoidant conflict management strategies at each of the six time points revealed complex but discernable patterns (Fig. 7.1). Each team is represented in a separate row in the figure, while the members' scores in cooperative, competitive, and avoidant conflict management strategies are presented across the three columns; the lines within each graph are data for individual members at the six measurement intervals.

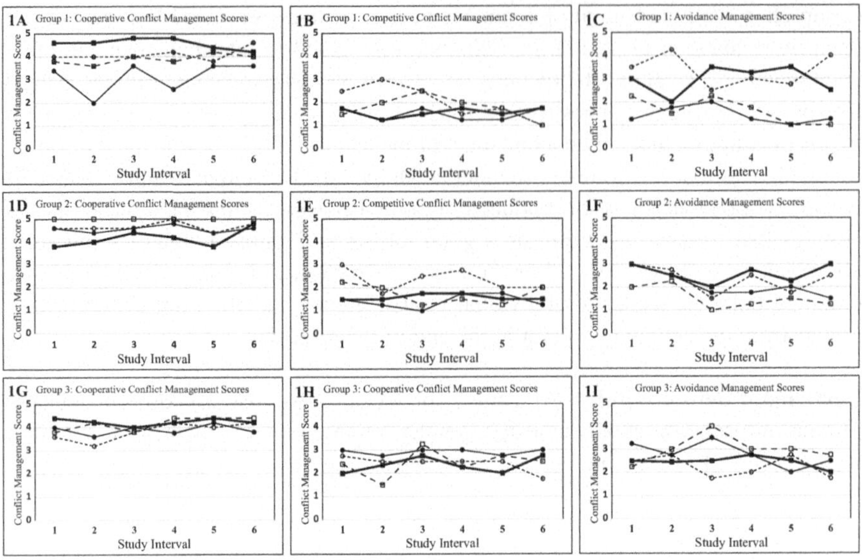

Fig. 7.1 Qualitative panel graphs showing members' temporal use of cooperative, competitive, and avoidant conflict management strategies across teams. (*Note:* Participants' scores in cooperative, competitive, and avoidant conflict management strategies (arranged in columns) at each of the six measurement intervals, organized based on team membership (arranged in rows))

Qualitatively, members of team 2 and team 3 shared the most similarity with their cooperative conflict management strategies, while those in team 1 had generally greater variability in individual members' use of the cooperative strategy (Fig. 7.1a–c). Interestingly, participants in team 1 exhibited the overall lowest amount of competitive conflict management strategy, especially later in the semester, when the team members' individual scores appeared to converge on what may be an attractor state (Fig. 7.1d). While team 2 exhibited similar levels of competitive conflict, those in this team did not appear to present a decline in this conflict management strategy during the later semester (Fig. 7.1e). Finally, it appears team 3 utilized the competitive conflict management approach more than the other teams, with the members having moderately high scores in this domain throughout the semester (Fig. 7.1f). In terms of the avoidant strategy, the only clearly evident distinction between the teams is found with team 1, where the members' reported use of the avoidant strategy varied greatly throughout the semester with divergence in members' use of this strategy late in the semester – two team members increasingly relied upon this particular strategy, while two others became less reliant upon the avoidant conflict management style (Fig. 7.1g). In the case of teams 2 and 3, moderate avoidant strategy scores were apparent with perhaps a very slight declining trend in members' use of this strategy (Fig. 7.1h, i). Considering that team 1 was the most successful with their project task, it is interesting that this team's members had the greatest variation in their use of cooperative conflict management. Although team

1's fairly low scores in competitive conflict management strategy might be predicted of a successful team, it is apparent that this team's success is related to more than this single factor. Indeed, the differences in the three teams' competitive conflict management strategy scores appear unlikely to be statistically significant, considering the variation in individual members' scores across the six intervals. The variation in the avoidant conflict management strategy use, especially with team 1, is also interesting and is a phenomenon that has not been extensively studied (Roloff and Ifert 2000; Tjosvold and Sun 2002). Building on these tentative results, we next applied a variety of inferential statistics to more conclusively discern the relationship between group success and the different patterns of cooperative, competitive, and avoidant conflict management strategies.

For our quantitative analysis, we initially applied average intraclass correlation coefficients to better understand whether team members' different conflict management strategies were, overall, more or less consistent during the duration of their semester-long project. The average intraclass correlation coefficient is used to assess score consistency rather than absolute agreement (see Shrout and Fleiss 1979). Interesting differences in the intraclass correlations between each team's conflict management styles were evident and suggestive of our core idea that the complex interaction between the different styles influences team performance (Table 7.1).

From a lay perspective, it might be presumed that the most successful teams have members who are more or less consistent in the different conflict management strategies they apply. Our preliminary results, however, reveal that this is not completely accurate. For example, team 1 concluded the semester with the most successful team project yet had the least consistency in members' reliance on a competitive conflict management approach ($ICC = 0.350$). In contrast, members in team 1 exhibited highly consistent scores in measures of their cooperative and avoidant strategy use ($ICC = 0.928$ and 0.929, respectively). The intraclass correlation coefficients, combined with our previous descriptive analysis of members' conflict management strategies over time, provide insight into the team's dynamics. Members of team 1 tended to report great reliance on a consistent cooperative conflict management approach. These members also had fairly consistent avoidant strategy use over time, qualitatively noting two members had a greater tendency to rely on this particular management strategy. While team 1 presented lower reported use of the competitive conflict management strategy, the members were not rigidly locked into this mode; instead, with the lower consistency in the competitive strategy scores per the

Table 7.1 Average intraclass correlation coefficients

	Average cooperation	Average competition	Average avoidance
Team 1	0.928**	0.350^{NS}	0.929**
Team 2	0.945**	0.865**	0.909**
Team 3	0.621^{NS}	0.669^{M}	0.639^{NS}

Note: ** significant at $p < 0.01$, *M* marginal significance, *NS* not significant.

intraclass correlation coefficient analysis, this team's members flexibly applied the competitive strategy. Considering Fig. 7.1b, this is most evident in how the team members experienced slightly greater perception of a competitive strategy early in their teamwork but later transitioned out of this conflict management approach.

Another assumption about conflict management in successful teams that might be made is that more successful teams tend to favor a cooperative approach to coordinating their actions. Interestingly, our study suggests that the relationship between conflict management approaches and teamwork outcomes is much more dynamic. Considering again our qualitative assessment, the overall highest level of reported cooperative conflict management strategy use occurred with members in team 2 (Fig. 7.1d). Furthermore, these team members had the most consistency across their cooperative strategy scores over time ($ICC = 0.945$). Despite consistently high levels of perceived cooperation, team 2 was only moderately successful. Following what other researchers have found in studies on team performance (e.g., Losada 1999), a harmonious interaction alone is not necessarily a characteristic of the best performing teams. Examination of team 2's avoidant and competitive management strategy data provides further preliminary insight. As with team 1, the members in team 2 had fairly consistent reported use of an avoidant strategy ($ICC = 0.909$). Within the intraclass correlations, however, an important difference is evident between these teams – team 2, which was less successful, had the highest consistency in their appeal to a competitive management strategy ($ICC = 0.865$). These initial results evoke a tantalizing hypothesis ripe for further study. It appears that a critical difference between highly successful and modestly performing teams may be how they utilize a competitive conflict management approach; the most successful teams have members who coordinate successfully and develop consistent ways of applying cooperative and avoidant conflict management strategies. However, these better performing teams remain flexible with how, and when, they apply competitive strategies. As a result, high-performance teams exhibit generally less consistency in competitive strategy reliance. When juxtaposed against more ordinary performing teams, the less successful teams share similar characteristics with members' adoption of cooperative and avoidant approaches; but, the key difference is that less successful teams over-coordinate and develop competitive management strategies that are more consistent, hence, less adaptive.

Finally, are there distinct characteristics of low-performance teams when considering conflict management dynamics? While it is a common belief that less successful teams have less cohesion among members and are more prone to intragroup competition, our data seem to present a richer insight. Across the three types of conflict management strategies, the third team in our study, which had the poorest project performance, was the only team whose members exhibited insignificant consistencies in use of any of the conflict management strategies (see Table 7.1). While the third team exhibited generally high levels of cooperative conflict management (Fig. 7.1g), the intraclass correlations among members' cooperative strategy scores were lower than other teams ($ICC = 0.621$) and failed to achieve statistical significance. In addition, their competition scores, while modest, were not especially higher than other teams' (Fig. 7.1h). By exploring our data from a dynamical systems-inspired perspective, we tentatively suggest that team members' tendency to use any particular combination of conflict management strategies alone is not linked to team outcomes in isolation. Instead, the type of strategies employed combined with whether the strategies are – or are not – utilized in a consistent way over time is what matters for the outcomes. Whereas the most successful team (one) and moderately successful team (two) exhibited high consistency in at least two types of conflict management strategies, the least successful team (three) failed to achieve consistency in any of the management strategies. The lack of consistency is particularly interesting from a dynamical systems perspective, as we elaborate on in our conclusion.

While the intraclass correlation coefficients provide tentative insight into the emergence of more, or less, consistent perceived use of the various conflict management strategies in teams who are working toward specific task completion, such results do not fully address the underlying team dynamics that may be pertinent to team success. Following key tenets from dynamical systems theory, we next applied discriminant analysis in a unique way to understand how the conflict management strategies unfolded in the teams over the course of their interactions. This approach further permitted assessment of whether the observed patterns in team conflict management were linked to internal events, external factors, or some combination.

Traditionally, discriminant analysis is a procedure used for discerning key quantitative characteristics that distinguish two or more categorical groups. The procedure thus confers predictive benefit, similar to regression techniques, in that it may help identify important group differences in multiple quantitative variables. In contrast with similar methods, such as cluster analysis, discriminant analysis is applicable when the group categories are already specified (see Kachigan 1991; Todman and Dugard 2007). With our study, our interest was in identifying whether differences emerged in cooperative, competitive, or avoidant conflict management strategies across the six intervals within each of the three teams. Since it was known which team was most (one) versus moderately (two) versus least (three) successful, it was plausible to use the team membership as a categorical variable for the discriminant analysis. Importantly, we applied this technique in a way that acknowledges the importance of time in the evolution of team interactions. Instead of simply performing a single discriminant analysis including team members' conflict management scores from all six intervals, we performed a series of *iterated discriminant analyses* in which we sequentially tested the success of the models to predict group membership based on the scores as the data accumulated across the study. That is, we initially performed the analysis considering only data from the first interval; next, we ran a new discriminant analysis including data at only the first two intervals and so forth until all six intervals were accounted for. This method, while novel, is partly inspired by autocorrelation.

For the iterated discriminant analysis, participants' cooperative, competitive, and avoidant strategy scores from each successive set of measurement intervals (intervals 1 through 6) were used for group classification using the stepwise method with an upper bound Fisher's statistic $p = 0.05$ for variable inclusion and a lower bound $p = 0.10$ for exclusion (see Table 7.2). In each case, the variable that minimized the

Table 7.2 Iterated discriminant analyses

		Study interval and variables included in discriminant analysis					
		1	2	3	4	5	6
Conflict management strategy	Cooperation	Omit	Omit	Omit	Omit	Omit	Include t = 6
	Competition	Omit	Omit	Include t = 3	Include t = 4	Include t = 5	Include t = 5
	Avoidance	Omit	Omit	Omit	Omit	Omit	Omit
	Accuracy	*NS*	*NS*	58.3%	58.3%	58.3%	83.3%

overall Wilks' lambda was entered. Using these general guidelines for each discriminant analysis, none of the conflict management strategies afforded significant correct classification of group measurement for scores recorded during the first two intervals. However, improved classification began once data from the third interval's iteration were included. When relying on the competitive conflict management scores from the third interval alone, 58.3 percent of the participants were correctly classified according to whether they were members of the highest-, middle-, or lowest-performing team. Interestingly, the classification was correct exclusively for the middle- and lowest-performing teams; members in the highest-performing team were classified incorrectly. These results remained unchanged when the fourth interval's conflict strategy scores were included – the fourth interval scores proved insignificant for classification purposes. When including the fifth interval, however, there was a 58.3 percent correct classification using the competitive scores from that interval. With the fifth interval data incorporated in the iterated discriminant analysis, the competitive scores from the third interval no longer achieved significance for classification purposes. Importantly, the fifth interval competitive strategy scores did correctly classify some members of the best performing team, whereas the third interval scores failed in this respect. Finally, significantly improved group classification occurred when the conflict management strategy scores from all six intervals were included in a single discriminant analysis. With such data, 83.3 percent of participants were correctly classified into either the highest-, middle-, or lowest-performing team. For this final analysis, only the fifth interval competitive strategy scores and sixth interval cooperative strategy scores were included in the classification.

The iterated discriminant analysis results reveal interesting patterns that provide preliminary insight into how the conflict management strategies are linked to team success. Since conflict management strategy data from the first two intervals were insignificant in the group membership classifications, it appears that early team interactions are not a primary driving factor in final performance outcomes. Rather, team responses to specific, external situational pressures may be more consequential. This tentative suggestion is made considering that the third interval coincided with when each team had a midsemester project update due. Although modest, a more competitive conflict management strategy among members (see Fig. 7.1) yielded a 58.3 percent correct classification of members in the two lower-performing teams. Evidently, reliance on a more competitive strategy at critical project junctures, such as around deadlines, is associated with subsequently poorer project outcomes. This interpretation makes sense considering that the competitive strategy scores again emerged as significant predictors of group membership at the fifth interval, which was close to the final project deadline. However, the best classification emerged when simultaneously accounting for the fifth interval competitive strategy scores and sixth interval cooperative strategy scores. Implicitly, whether more (or less) competitive conflict management strategies are employed during critical project junctures is related to team success. Furthermore, it may be that teams that better manage and limit use of the competitive strategy may adopt more cooperative strategies as their work draws to a close.

7.6 Discussion

The overall objective of this chapter was to present and discuss the relevance and application of the dynamical systems framework in the study of conflict management processes in teams. In particular, we were interested in addressing three key research questions: first, do individual team members' conflict management strategies – cooperative, competitive, or avoiding – combine in discernible patterns that contribute to significant differences in their teams' overall effectiveness; second, can fundamental properties of dynamical systems, such as self-organization leading to emergent phenomenon and attractor states, be observed in team conflict management dynamics; and third, how might internal dynamics and consequent outcomes of a team conflict management system be affected by external influences or contextual conditions? In the following paragraphs, we summarize tentative observations and findings drawn from our research thus far, discuss implications and limitations associated with our preliminary study, as well as propose promising directions for future inquiry.

7.6.1 Tentative Observations and Findings

To explore our first research question, that is, whether meaningful patterns of internal conflict management dynamics can be observed in ways that would distinguish high-performing teams from lower-performing ones, we conducted a qualitative assessment using panel graphs that depicted the temporal patterns of members' conflict management scores across the three teams and a quantitative analysis using intraclass correlation indices. Overall, our qualitative assessment provided insight about the different team dynamics that were consistent with the quantitative results.

Based on our examination of the qualitative panel graphs, we observed that members of the highest-performing team (1) used high yet variable levels of cooperative strategies over time, alongside a generally low and steadily declining pattern of competitive strategies. Moreover, the use of avoidant strategies among team 1 members revealed a divergence or "split" over time: two members were consistently higher in their use of avoidant strategies, while the other two members in the team were consistently lower. As for moderately performing team 2, members were consistent in relying on highly cooperative strategies and limiting their use of competitive strategies over time, although the limited competitive strategy application may not have been completely beneficial. Finally, for members in lowest-performing team 3, their use of cooperative strategies was less consistent than observed in teams 1 and 2.

Along a similar vein, our quantitative analysis also revealed that the highest-performing team (1) relied most on cooperative and avoidant approaches, but least on competitive approaches, compared to the moderately high-performing team (2), which engaged in consistently high levels of cooperative, avoidant, *and* competitive approaches. As for the lowest-performing team (3) in our study, its members did not use consistent cooperative, competitive, or avoidant strategies when handling conflict. These results also seem to be in line with past research findings that higher-performing teams, on average, tend to engage in highly consistent cooperative strategies, compared to lower-performing teams (e.g., Alper et al. 2000; Tjosvold et al. 2006).

Taken altogether, the patterns revealed in our qualitative and quantitative analyses suggest that higher-performing teams are likely to be more adaptive in their use of different conflict management strategies, compared to lower-performing teams. More importantly, this adaptive ability in using different conflict management strategies among high-performing teams may be especially contingent upon their flexible use of competitive strategies when addressing conflict.

In addressing our second question on whether dynamical properties, such as time evolution, self-organization, and attractor states, could be observed in team conflict management, we also relied on the qualitative panel graphs and intraclass correlational indices to make some further intriguing and interesting observations. To begin, the variability in the cooperative and avoidant strategies observed in team 1, based on the qualitative graphs, was not without pattern. As implied by relatively high intraclass correlation coefficients in our quantitative analysis, members in this best performing team seemed to exhibit a flexible attractor-like dynamic with their adoption of these strategies. In contrast, the low coefficients found with these team members' competitive strategies implied the steady decline occurred as a result of absent emergent attractor dynamics in this particular mode of management strategy.

From the intraclass correlation coefficient results, we also noted that the moderately performing team (two) seems to have developed an attractor dynamic governing their competitive strategy. With this strategy seemingly forced to remain fairly low across the team project duration, perhaps this team did not experience sufficient flexibility or tension in their interactions to facilitate the necessary energy for stronger performance (cf. Losada 1999).

As for the lowest-performing team (3), it had the lowest intraclass correlation coefficient for cooperative strategy consistency over time, which implies that it never developed an attractor state for this conflict management strategy. For this team, the absence of a conflict management strategy attractor may have left them unable to effectively coordinate. This greatly contrasts with the better performing teams and suggests an absence of a consistent cooperative approach may be a critical characteristic of a team that is in danger of nonoptimal performance. While tentative, this interpretation makes sense considering Losada's (1999) team performance research as well as in light of other research on group coordination and interpersonal synchrony (e.g., Vallacher et al. 2005). This finding is particularly interesting as it suggests the level of cooperation may not be the most decisive factor in team success. Instead, team success may be more contingent upon group dynamics. Teams that are energized by their tensions and can set these aside to build more consistent cooperative and avoidant strategies for managing their conflict may be those capable of the highest performance.

Our results are of further interest considering the apparent differences in the degree to which teams relied on the different strategies. While studies have shown some degree of tension is worthwhile for effective team performance (e.g., Losada 1999), the quality of that tension seems to matter and distinguish the best performing teams. We identified that the best performing team in our study exhibited inconsistent reliance upon a competitive strategy with greater competitive strategy use early in the interaction. In contrast, the next best performing team (two) had the highest consistency in applying a competitive strategy to their group interactions, but this consistency was linked to low reliance on this strategy. From a dynamical systems perspective, these differences may reveal that the top performing teams tend to develop consistent patterns of using cooperative and avoidant conflict management strategies, which would suggest these function more or less as an attractor. In contrast, the competitive strategy does not emerge as an attractor within these high-performance team members' interactions. It would be incorrect to suggest that this means high-performance teams do not exhibit competitive strategies. Indeed, from the team (one) members' individual data (Fig. 7.1), moderately competitive strategies are evident, especially early in the group interaction (intervals 1 through 3). What appears to happen is that high-performing teams explore competitive strategies as a form of energizing tension early in their interactions, but as the group gains cohesion, such strategies may fall aside to cooperative and avoidant approaches. While this possibility warrants further empirical investigation, it makes sense from a dynamical systems perspective, for if both the cooperative and avoidant strategies develop attractor properties, these may become more automatically utilized over a competitive style in these groups. Furthermore, this preliminary assessment fits with general ideas about social evolution where, in general, people tend to more often rely upon cooperative reciprocity over an antagonistic exchange (e.g., Axelrod and Hamilton 1981; Nowak 2006).

The third and last key question we sought to explore in this preliminary research has to do with whether the patterns in how high- and low-performing teams adopted cooperative, competitive, and avoiding strategies differ at critical junctures during their team life spans. As discussed earlier in the chapter, all teams had to complete a midpoint project update (i.e., a critical juncture) at Time 3 or halfway through the study. Based on our iterated discriminant analyses, it seems that the emergence of an attractor-like state in conflict management strategies may be pertinent at critical junctures in a team project. For example, the more consistent styles employed by the higher-performing teams are evident in their data exhibiting higher intraclass correlations over time. Thus, the stable – yet flexible – approach to managing conflict among the group members may have facilitated better responses to those periods when major project deadlines were imminent (this, identified through the intervals for which the discriminant analysis yielded significant prediction for team membership at the project midpoint and toward the project final deadline). Put otherwise, we found that teams that relied more on competitive strategies around critical junctures, such as the midpoint project update or toward the end of their project, performed more poorly than those who were able to limit their use of competitive strategies or maintain more cooperative strategies during these critical points in time. Although tentative, our results do seem to demonstrate the importance of external influences or events in shaping teams' responsiveness or adaptive ability in adopting appropriate conflict management strategies over time.

7.6.2 Implications, Limitations, and Future Directions

Our research contributes to a richer understanding of how successful teams emerge from the complex and dynamic interaction between various internal and external factors over time. By considering how the different conflict management strategies evolved over time, we note that the fundamental difference between the best and lesser performing teams appears to be in whether a consistent (attractor-like) adoption of cooperative and avoidant strategies takes place over a competitive strategy. Consistency is not the sole key characteristic of the best performing teams, however, nor is the limitation of competitive conflict management strategies. As shown in our data, a high-performing team is open to competitive strategies during the formative stages of the members' work together. However, an important transition takes place over time as the competitive strategies are inconsistently used and fall aside for members' preference to instead adopt cooperative or avoidant approaches to managing the interaction. These novel findings fit with some ideas from team performance research, where high-performance teams tend to have an energy and flexibility in their interactions (e.g., Fredrickson and Losada 2005; Losada 1999) and general findings that find flexibility is linked to more optimal functioning in numerous social and psychological domains (e.g., Bartoli et al. 2010; Isbell et al. 2013; Kashdan and Rottenberg 2010; Michaels et al. 2013b).

Flexibility may be essential for groups to avoid stagnation in their interactions and instead maintain energy. Indeed, for the best performing team in our study, this energy may be linked to tensions aroused by early use of competitive strategies that then gets transferred to enthusiasm for the interaction as cooperation takes root. In the case of poorer-performing teams, it is possible that this energizing tension never takes root. As our study's preliminary results demonstrate, a poor-performing team might not be dominated by competitive interactions among members. Instead, the members may seek to limit and avoid competitive conflict management strategies. Over time, the avoidance of such strategies may lead to an attractor-like state constraining the group during times of tension, leading to inability to flexibly cope with external pressures. This theory fits with our identifying that the poorest-performing team's competitive strategy scores increased around project deadlines.

Dynamically considered, future work might explore whether avoidance of a competitive strategy leads to an underlying latent attractor for conflict; that is, by avoiding the competitive strategy, do the group members inadvertently set themselves up for greater conflict when external pressures arise and lead to strain in their use of inconsistent cooperative conflict management? If the cooperative strategy lacks coherence through inconsistent application, then there is a real possibility that such external pressures could force the group out of their comfort zone, leading to new and unexpected dynamics. These hypothetical possibilities fit with many general ideas from dynamic systems psychology (Nowak and Vallacher 1998) and empirical studies that find psychologically seeking to avoid something often makes it more likely that whatever one avoids becomes consciously salient and thus more prone to influence behavior (e.g., Wegner et al. 1987, 1991).

Extending our research to application, our exploratory study advocates team success, or lack thereof, is not linked to simple causality. Rather, understanding team dynamics and how teams evolve over time is fundamental for gaining an understanding of whether a team is on a trajectory for excellent performance or is potentially progressing toward nonoptimal outcomes. While our results require further empirical studies and replication, they preliminarily indicate that limiting a competitive conflict management strategy early in a group's interaction may not cultivate a high-performance team. Instead, early use of a competitive strategy seems as important as early adoption of a consistent cooperative conflict management approach for the iterative progression toward high performance. Competition can be good if it does not emerge as the predominant mode of conflict management.

Furthermore, over time, the best performing teams seem to have an emerging stable reliance on cooperative conflict management strategies. The consistent reliance on this strategy seems especially important for periods of external pressure, as the external perturbation from a project deadline or unanticipated problem will tend to exacerbate existing group tensions surrounding interpersonal dynamics between members and individual factors within the members (e.g., proneness to stress or anxiety, personal life problems). An absence of a consistent cooperative conflict management strategy may lead to less stable, more conflict-ridden, confused interactions within the group in times of external pressure. In contrast, if consistent cooperative conflict management strategies are in place through the group's own self-organization, then work pressures may cause tension within the group, but the prevailing cooperative strategy can allow the work members to cohesively cooperate and harness the tension to accomplish the necessary goals related to their work together. We look forward to further examining these possible applications of our work with future research.

While our preliminary findings provided some promising insights into the study of team conflict management from the dynamical systems perspective, we are also mindful of the limitations associated with the research discussed in this chapter. Most notably, the findings presented here were based on a relatively small team sample, and so, continuing to increase the team sample size would be important in strengthening the tentative conclusions drawn thus far. Of course, our conclusions are also based on a sample taken from a single location and from teams working on a class project. For better establishing external validity and ensuring our conclusions are generalizable, further replications are necessary from a variety of teams in different contexts. There may, for example, be interesting differences between teams depending on whether the teams' goals are tied to intrinsic versus extrinsic motives (e.g., Higgins 2007) or carry outcomes that have heavy consequences.

Further investigations in this area would also benefit from using alternative research designs and methods, such as daily experience sampling (e.g., Sherman et al. 2015; Spain et al. 2000) and dynamic systems techniques that better capture higher-resolution patterns of data over time, such as the mouse paradigm method (e.g., Michaels et al. 2013a; Vallacher et al. 2002a). We would also encourage future researchers to augment existing findings with mathematical modeling or computational simulations (e.g., Liebovitch et al. 2010; Losada 1999; Nowak et al. 1994; Sun 2008) of team conflict management dynamics that might allow us to make informed predictions about how the dynamical patterns of conflict management in teams may change and evolve. Considering the complexity involved in group processes like team development and performance, perhaps the best approach would be to utilize all of these (and other) methods to develop a deeper understanding of the various inter- and intrapersonal processes and their interactions. A programmatic interdisciplinary approach would thus maximize the potential for insightful discoveries in this exciting domain of research that may yield beneficial applications for team performance in a variety of situational contexts.

7.7 Conclusion

Society is built upon the capacity to form effective teams that are capable of combining individuals' physical and mental capabilities to solve problems, forge common goals, and successfully adapt to the demands of an often challenging world (cf. Nowak 2006). Despite the significant part teams play in much of people's daily life, exactly what factors facilitate formation of a highly effective team versus a team that is mired in petty conflict or uncreative solutions remains at least somewhat mysterious. We suggest that a dynamical systems approach to studying how teams form, evolve, and realize goal outcomes is an essential iteration in the research to unlock the secrets of what allows some teams to excel, even in demanding situations. While the study we describe in this chapter is preliminary, it is a first step in this progression toward better understanding the time-dependent evolution of excelling versus underperforming teams. As our future studies, and hopefully those others might conduct if they are inspired to embrace dynamical systems theory and methodology, lead to new results, perhaps the findings will allow us to create interventions that help people form stronger, more effective teams in the government, the workplace, and educational settings. Regardless of what future discoveries reveal, research on team effectiveness promises to be dynamic and exciting for decades to come.

References

Alper, S., Tjosvold, D., & Law, K. S. (2000). Conflict management, efficacy, and performance in organizational teams. *Personnel Psychology, 53*(3), 625–642. https://doi.org/10.1111/j.1744-6570.2000.tb00216.x

Amason, A. C., & Schweiger, D. M. (1997). The effects of conflict on strategic decision making effectiveness and organizational performance. In C. K. W. De Dreu & E. Van de Vliert (Eds.), *Using conflict in organizations* (pp. 101–115). Beverly Hills: Sage.

Amason, A. C., Thompson, K. R., Hochwarter, W. A., & Harrison, A. W. (1995). Conflict: An important dimension in successful management teams. *Organizational Dynamics, 24*(2), 20–35. Retrieved from http://www.sciencedirect.com/science/article/B6W6S-4B7GPCN-2R/2/65618710490c5f7231e9a3eaf380a332

Avgar, A. C., & Neuman, E. J. (2015). Seeing conflict: A study of conflict accuracy in work teams. *Negotiation and Conflict Management Research, 8*(2), 65–84. https://doi.org/10.1111/ncmr.12048

Axelrod, R., & Hamilton, W. D. (1981). The evolution of cooperation. *Science, 211*, 1390–1396 https://doi.org/10.1126/science.7466396

Bartoli, A., Bui-Wrzosinska, L., & Nowak, A. (2010). Peace is in movement: A dynamical systems perspective on the emergence of peace in Mozambique. *Peace and Conflict: The Journal of Peace Psychology, 16*, 211–230 https://doi.org/10.1080/10781911003691609

Bear, J. B., Weingart, L. R., & Todorova, G. (2014). Gender and the emotional experience of relationship conflict: The differential effectiveness of avoidant conflict management. *Negotiation and Conflict Management Research, 7*(4), 213–231. https://doi.org/10.1111/ncmr.12039

Behfar, K. J., Peterson, R. S., Mannix, E. a., & Trochim, W. M. K. (2008). The critical role of conflict resolution in teams: A close look at the links between conflict type, conflict management strategies, and team outcomes. *The Journal of Applied Psychology, 93*(1), 170–188. https://doi.org/10.1037/0021-9010.93.1.170

Behfar, K., Friedman, R., & Brett, J. (2016). Managing co-occurring conflicts in teams. *Group Decision and Negotiation, 25*(3), 501–536. https://doi.org/10.1007/s10726-015-9450-x

Boros, S., Meslec, N., Curseu, P. L., & Emons, W. (2010). Struggles for cooperation: Conflict resolution strategies in multicultural groups. *Journal of Managerial Psychology, 25*(5), 539–554. https://doi.org/10.1108/02683941011048418

Carnevale, P. J., & Probst, T. M. (1998). Social values and social conflict in creative problem solving and categorization. *Journal of Personality and Social Psychology, 74*(5), 1300–1309. https://doi.org/10.1037/0022-3514.74.5.1300

Chen, G., & Tjosvold, D. (2002). Conflict management and team effectiveness in China: The mediating role of justice. *Asia Pacific Journal of Management, 19*, 557–572. Retrieved from http://www.springerlink.com/index/V24324R4X35776M0.pdf

Chen, G., Liu, C., & Tjosvold, D. (2005). Conflict management for effective top management teams and innovation in China. *Journal of Management Studies*. https://doi.org/10.1111/j.1467-6486.2005.00497.x

Coleman, P. T. (2011). *The five percent: Finding solutions to seemingly impossible conflicts*. New York: Public Affairs.

Coleman, P. T., Vallacher, R. R., Nowak, A., & Bui-Wrzosinska, L. (2007). Intractable conflict as an attractor: A dynamical systems approach to conflict escalation and intractability. *American Behavioral Scientist, 50*(11), 1454–1475. https://doi.org/10.1177/0002764207302463

De Dreu, C. K. W. (2006). When too little or too much hurts: Evidence for a curvilinear relationship between task conflict and innovation in teams. *Journal of Management, 32*(1), 83–107. https://doi.org/10.1177/0149206305277795

De Dreu, C. K. W., & Van Vianen, A. E. M. (2001). Managing relationship conflict and the effectiveness of organizational teams. *Journal of Organizational Behavior, 22*(3), 309–328. https://doi.org/10.1002/job.71

De Dreu, C. K. W., & Weingart, L. R. (2003). Task versus relationship conflict, team performance, and team member satisfaction: A meta-analysis. *Journal of Applied Psychology, 88*(4), 741–749. https://doi.org/10.5465/APBPP.2002.7516590

De Dreu, C. K. W., & West, M. A. (2001). Minority dissent and team innovation: The importance of participation in decision making. *Journal of Applied Psychology, 86*(6), 1191–1201. https://doi.org/10.1037/0021-9010.86.6.1191

de Wit, F. R. C., Greer, L. L., & Jehn, K. A. (2012). The paradox of intragroup conflict: A meta-analysis. *Journal of Applied Psychology, 97*(2), 360–390. https://doi.org/10.1037/a0024844

DeChurch, L. A., & Marks, M. A. (2001). Maximizing the benefits of task conflict: The role of conflict management. *International Journal of Conflict Management, 12*(1), 4–22. https://doi.org/10.1108/eb022847

DeChurch, L. A., Hamilton, K. L., & Haas, C. (2007). Effects of conflict management strategies on perceptions of intragroup conflict. *Group Dynamics: Theory, Research, and Practice, 11*(1), 66–78. https://doi.org/10.1037/1089-2699.11.1.66

DeChurch, L. A., Mesmer-Magnus, J. R., & Doty, D. (2013). Moving beyond relationship and task conflict: Toward a process-state perspective. *Journal of Applied Psychology, 98*(4), 559–578. https://doi.org/10.1037/a0032896

Desivilya, H. S., & Yagil, D. (2005). The role of emotions in conflict management: The case of work teams. *International Journal of Conflict Management, 16*(1), 55–69. https://doi.org/10.1108/eb022923

Deutsch, M. (1949). A theory of co-operation and competition. *Human Relations, 2*(2), 129–152. https://doi.org/10.1177/001872674900200204

Deutsch, M. (1973). *The resolution of conflict: Constructive and destructive processes*. New Haven: Yale University Press.

Deutsch, M. (2014). Cooperation, competition, and conflict. In P. T. Coleman, M. Deutsch, E. C. Marcus, P. T. Coleman, M. Deutsch, & E. C. Marcus (Eds.), *The handbook of conflict resolution: Theory and practice* (3rd ed., pp. 3–28). San Francisco: Jossey-Bass.

Farmer, S. M., & Roth, J. (1998). Conflict-handling behavior in work groups: Effects of group structure, decision processes, and time. *Small Group Research, 29*(6), 669–713. https://doi.org/10.1177/1046496498296002

Fredrickson, B. L., & Losada, M. F. (2005). Positive affect and the complex dynamics of human flourishing. *American Psychologist, 60*, 678–686 https://doi.org/10.1037/0003-066X.60.7.678

Guastello, S., Koopmans, M., & Pincus, D. (Eds.). (2009). *Chaos and complexity in psychology: The theory of nonlinear dynamical systems*. New York: Cambridge University Press.

Hempel, P. S., Zhang, Z., & Tjosvold, D. (2008). Conflict management between and within teams for trusting relationships and performance in China, (April 2007). https://doi.org/10.1002/job

Higgins, E. T. (2007). Value. In A. W. Kruglanski & E. T. Higgins (Eds.), *Social psychology: Handbook of basic principles* (2nd ed., pp. 454–472). New York: The Guilford Press.

Holland, J. H. (1995). *Emergence: From chaos to order*. Reading, MA: Addison-Wesley.

Isbell, L. M., Lair, E. C., & Rovenpor, D. R. (2013). Affect-as-information about processing styles: A cognitive malleability approach. *Social and Personality Psychology Compass, 7*(2), 93–114 https://doi.org/10.1111/spc3.12010

Jarboe, S. C., & Witteman, H. R. (1996). Intragroup conflict management in task-oriented groups: The influence of problem sources and problem analyses. *Small Group Research, 27*(2), 316–338. https://doi.org/10.1177/1046496496272007

Jehn, K. A. (1994). Enhancing effectiveness: An investigation of advantages and disadvantages of value-based intragroup conflict. *International Journal of Conflict Management, 5*(3), 223–238. https://doi.org/10.1108/eb022744

Jehn, K. (1995). A multimethod examination of the benefits and detriments of intragroup conflict. *Administrative Science Quarterly, 40*(2), 256–282. https://doi.org/10.2307/2393638

Jehn, K. a. (1997). A qualitative analysis of conflict types and dimensions in organizational groups. *Administrative Science Quarterly, 42*(3), 530–557. https://doi.org/10.2307/2393737

Jehn, K. A., & Bendersky, C. (2003). Intragroup conflict in organizations: A contingency perspective on the conflict-outcome relationship. In B. Staw (Ed.), *Research in organizational behavior* (Vol. 25, pp. 187–242). https://doi.org/10.1016/S0191-3085(03)25005-X

Jehn, K. A., Greer, L., Levine, S., & Szulanski, G. (2008). The effects of conflict types, dimensions, and emergent states on group outcomes. *Group Decision and Negotiation, 17*(6), 465–495. https://doi.org/10.1007/s10726-008-9107-0

Jehn, K. A., Rispens, S., & Thatcher, S. M. B. (2010). The effects of conflict asymmetry on work group and individual outcomes. *Academy of Management Journal, 53*(3), 596–616. https://doi.org/10.5465/AMJ.2010.51468978

Johnson, D. W., & Johnson, R. T. (2005). New developments in social interdependence theory. *Genetic, Social, and General Psychology Monographs, 131*(4), 285–358. https://doi.org/10.3200/MONO.131.4.285-358

Johnson, D. W., & Johnson, R. T. (2011). Conflict in the classroom: Controversy and learning. *Review of Educational Research, 49*(1), 51–69 https://doi.org/10.3102/00346543049001051

Johnson, S. L., & Nowak, A. (2002). Dynamical patterns in bipolar depression. *Personality and Social Psychology Review, 6*, 380–387 https://doi.org/10.1207/S15327957PSPR0604_12

Jones, R. E., & White, C. S. (1985). Relationships among personality, conflict resolution styles, and task effectiveness. *Group & Organization Management, 10*(2), 152–167. https://doi.org/10.1177/105960118501000204

Jordan, P. J., & Troth, A. C. (2004). Managing emotions during team problem solving: Emotional intelligence and conflict resolution. *Human Performance, 17*(2), 195–218. https://doi.org/10.1207/s15327043hup1702_4

Kachigan, S. K. (1991). *Multivariate statistical analysis: A conceptual introduction* (2nd ed.). New York: Radius Press.

Kashdan, T. B., & Rottenberg, J. (2010). Psychological flexibility as a fundamental aspect of health. *Clinical Psychology Review, 30*, 467–480 https://doi.org/10.1016/j.cpr.2010.03.001

Kleinman, G., Palmon, D., & Lee, P. (2003). The effects of personal and group level factors on the outcomes of simulate auditor and client teams. *Group Decision and Negotiation, 12*(1), 57–84 https://doi.org/10.1023/A:1022256730300

Kuhn, T., & Poole, M. S. (2000). Do conflict management styles affect group decision making? *Human Communication Research, 26*(4), 558–590. https://doi.org/10.1111/j.1468-2958.2000.tb00769.x

Larsen, R. J., & Diener, E. (1985). A multitrait-multimethod examination of affect structure: Hedonic level and emotional intensity. *Personality and Individual Differences, 6*, 631–636 https://doi.org/10.1016/0191-8869(85)90013-3

Liebovitch, L. S., Vallacher, R. R., & Michaels, J. L. (2010). Dynamics of cooperation-competition models. *Peace and Conflict: Journal of Peace Psychology, 16*, 175–188 https://doi.org/10.1080/10781911003691625

Liu, J., Fu, P., & Liu, S. (2009). Conflicts in top management teams and team/firm outcomes: The moderating effects of conflict-handling approaches. *International Journal of Conflict Management, 20*(3), 228–250. https://doi.org/10.1108/10444060910974867

Losada, M. (1999). The complex dynamics of high performance teams. *Mathematical and computer modeling, 30*, 179–192. https://doi.org/10.1016/S0895-7177(99)00189-2

Marks, M. A., Mathieu, J. E., & Zaccaro, S. J. (2001). A temporally based framework and taxonomy of team processes. *Academy of Management Review, 26*(3), 356–376. https://doi.org/10.5465/AMR.2001.4845785

Michaels, J. L., Vallacher, R. R., & Liebovitch, L. S. (2013). Volatile psychological dynamics in social interactions: Attitudes and emotions react asymmetrically to interaction shifts between agreement and disagreement. *Social Psychological and Personality Science, 4*, 705–713 https://doi.org/10.1177/1948550613482985

Michaels, J. L., Parkin, S. S., & Vallacher, R. R. (2013b). Destiny is in the details: Action identification in the construction and deconstruction of meaning in life. In J. A. Hicks & C. Routledge (Eds.), *The experience of meaning in life: Classical perspectives, emerging themes, and controversies* (pp. 103–116). New York: Springer.

Nemeth, C. (1995). Dissent as driving cognition, attitudes and judgments. *Social Cognition, 13*, 273–291 https://doi.org/10.1521/soco.1995.13.3.273

Nowak, M. A. (2006). Five rules for the evolution of cooperation. *Science, 314*, 1560–1563 https://doi.org/10.1126/science.1133755

Nowak, A., & Vallacher, R. R. (1998). *Dynamical social psychology*. New York: The Guilford Press.

Nowak, A., Vallacher, R. R., & Lewenstein, M. (1994). Toward a dynamic social psychology. In R. R. Vallacher & A. Nowak (Eds.), *Dynamical systems in social psychology* (pp. 279–293). San Diego: Academic Press.

Nowak, A., Vallacher, R. R., & Miller, M. E. (2003). Social influence and group dynamics. In T. Millon & M. J. Lerner (Eds.), *Handbook of psychology, volume 5: Personality and social psychology*. New York: Wiley.

Nowak, A., Deutsch, M., Bartkowski, W., & Solomon, S. (2010). From crude law to civil relations: The dynamics and potential resolution of intractable conflict. *Peace and Conflict: Journal of Peace Psychology, 16*, 189–209 https://doi.org/10.2139/ssrn.1493525

O'Neill, T. A., & Allen, N. J. (2014). Team task conflict resolution: An examination of its linkages to team personality composition and team effectiveness outcomes. *Group Dynamics: Theory, Research, and Practice, 18*(2), 159–173. https://doi.org/10.1037/gdn0000004

O'Neill, T. A., Allen, N. J., & Hastings, S. E. (2013). Examining the 'pros' and 'cons' of team conflict: A team-level meta-analysis of task, relationship, and process conflict. *Human Performance, 26*(3), 236–260 https://doi.org/10.1080/08959285.2013.795573

Oetzel, J. G. (1999). The influence of situational features on perceived conflict styles and self-construals in work groups. *International Journal of Intercultural Relations, 23*(4), 679–695 https://doi.org/10.1016/S0147-1767(99)00015-2

Ohland, M. W., Loughry, M. L., Woehr, D. J., Bullard, L. G., Felder, R. M., Finelli, C. J., et al. (2012). The comprehensive assessment of team member development of a behaviorally anchored rating scale for self- and peer evaluation. *Academy of Management Learning and Education, 11*(4), 609–631. https://doi.org/10.5465/amle.2010.0177

Okhuysen, G. A., & Richardson, H. A. (2007). Group conflict as an emergent state: Temporal issues in the conceptualization and measurement of disagreement. In K. J. Behfar & L. L. Thompson (Eds.), *Conflict in organizational groups: New directions in theory and practice* (pp. 145–179). London: Kogan Page.

Paul, S., Samarah, I. M., Seetharaman, P., & Mykytyn Jr., P. P. (2005). An empirical investigation of collaborative conflict management style in group support system-based global virtual teams. *Journal of Management Information Systems, 21*(3), 185–222 https://doi.org/10.1080/07421222.2004.11045809

Polzer, J. T., Milton, L. P., & Swann Jr., W. B. (2002). Capitalizing on diversity: Interpersonal congruence in small work groups. *Administrative Science Quarterly, 47*, 296–324 https://doi.org/10.2307/3094807

Poole, M. S., & Dobosh, M. (2010). Exploring conflict management processes in jury deliberations through interaction analysis. *Small Group Research, 41*(4), 408–426. https://doi.org/10.1177/1046496410366310

Rahim, M. A., & Bonoma, T. V. (1979). Managing organizational conflict – model for diagnosis and intervention. *Psychological Reports, 44*(3), 1323–1344 https://doi.org/10.2466/pr0.1979.44.3c.1323

Rispens, S., Greer, L. L., & Jehn, K. a. (2007). It could be worse: A study on the alleviating roles of trust and connectedness in intragroup conflicts. *International Journal of Conflict Management, 18*(4), 325–344. https://doi.org/10.1108/10444060710833450

Roloff, M. E., & Ifert, D. E. (2000). Conflict management through avoidance: Withholding complaints, suppressing arguments, and declaring topics taboo. In *Balancing the secrets of private disclosures* (pp. 151–163). Mahwah: Lawrence Erlbaum Associates Publishers.

Schulz-Hardt, S., Jochims, M., & Frey, D. (2002). Productive conflict in group decision making: Genuine and contrived dissent as strategies to counteract biased information seeking. *Organizational Behavior and Human Decision Processes, 88*, 563–586 https://doi.org/10.1016/S0749-5978(02)00001-8

Sherman, R. A., Rauthmann, J. F., Brown, N., Serfass, D. G., & Jones, A. B. (2015). The independent effects of personality and situations onreal-time expressions of behavior and emotion. *Journal of Personality and Social Psychology, 109*, 872–888 https://doi.org/10.1037/pspp0000036

Shrout, P. E., & Fleiss, J. L. (1979). Intraclass correlations: Uses in assessing rater reliability. *Psychological Bulletin, 86*(2), 420–428 https://doi.org/10.1037/0033-2909.86.2.420

Somech, A. (2008). Managing conflict in school teams: The impact of task and goal interdependence on conflict management and team effectiveness. *Educational Administration Quarterly, 44*(3), 359–390. https://doi.org/10.1177/0013161X08318957

Spain, J. S., Eaton, L. G., & Funder, D. C. (2000). Perspective on personality: The relative accuracy of self versus others for the prediction of emotion and behavior. *Journal of Personality, 68*, 837–867 https://doi.org/10.1111/1467-6494.00118

Staw, B. M., Sandelands, L. E., & Dutton, J. E. (1981). Threat- rigidity effects in organizational-behavior: A multilevel analysis. *Administrative Science Quarterly, 26*, 501–524 https://doi.org/10.2307/2392337

Strogatz, S. (2003). *Sync: The emerging science of spontaneous order*. New York: Hyperion Books.

Sun, R. (2008). *The Cambridge handbook of computational psychology*. New York: Cambridge University Press.

Thomas, K. W., & Kilmann, R. H. (1974). *The Thomas-Kilmann conflict mode instrument*. Tuxedo: Xicom.

Tjosvold, D. (1984). Effects of crisis orientation on managers' approach to controversy in decision making. *Academy of Management Journal, 27*, 130–138 https://doi.org/10.5465/255961

Tjosvold, D. (1990). Flight crew collaboration to manage safety risks. *Group & Organization Management, 15*(2), 177–191. https://doi.org/10.1177/105960119001500204

Tjosvold, D. (1998). Cooperative and competitive goal approach to conflict: Accomplishments and challenges. *Applied Psychology, 47*(3), 285–313. https://doi.org/10.1111/j.1464-0597.1998.tb00025.x

Tjosvold, D. (2006). Defining conflict and making choices about its management. *International Journal of Conflict Management, 17*(2), 87–95. https://doi.org/10.1108/10444060610736585

Tjosvold, D. (2008). The conflict-positive organization: It depends upon us. *Journal of Organizational Behavior, 29*(1), 19–28. https://doi.org/10.1002/job.473

Tjosvold, D., & Sun, H. F. (2002). Understanding conflict avoidance: Relationship, motivations, actions, and consequences. *International Journal of Conflict Management*. https://doi.org/10.1108/eb022872

Tjosvold, D., Law, K. S., & Sun, H. (2006). Effectiveness of Chinese teams: The role of conflict types and conflict management approaches. *Management and Organization Review, 2*(2), 231–252. https://doi.org/10.1111/j.1740-8784.2006.00040.x

Tjosvold, D., Yu, Z., & Wu, P. (2009). Empowering individuals for team innovation in China: Conflict management and problem solving. *Negotiation and Conflict Management Research, 2*(2), 185–206. https://doi.org/10.1111/j.1750-4716.2009.00036.x

Tjosvold, D., Wong, A. S. H., & Feng Chen, N. Y. (2014). Constructively managing conflicts in organizations. *Annual Review of Organizational Psychology and Organizational Behavior, 1*(1), 545–568. https://doi.org/10.1146/annurev-orgpsych-031413-091306

Todman, J., & Dugard, P. (2007). *Approaching multivariate analysis: An introduction for psychology*. New York: Psychology Press.

Vallacher, R. R., & Nowak, A. (2007). Dynamical social psychology: Finding order in the flow of human experience. In A. W. Kruglanski & E. T. Higgins (Eds.), *Social psychology: Handbook of basic principles* (2nd ed., pp. 734–758). New York: Guilford Press.

Vallacher, R. R., Nowak, A., Froelich, M., & Rockoff, M. (2002). The dynamics of self-evaluation. *Personality and Social Psychology Review, 6*, 370–379 https://doi.org/10.1207/S15327957PSPR0604_11

Vallacher, R. R., Read, S. J., & Nowak, A. (2002b). The dynamical perspective in personality and social psychology. *Personality and Social Psychology Review, 6*(4), 264–273. https://doi.org/10.1207/S15327957PSPR0604_01

Vallacher, R. R., Nowak, A., & Zochowski, M. (2005). Dynamics of social coordination: The synchronization of internal states in close relationships. *Interaction Studies, 6*, 35–52.

Vallacher, R. R., Coleman, P. T., Nowak, A., & Bui-Wrzosinska, L. (2010). Rethinking intractable conflict: The perspective of dynamical systems. *The American Psychologist, 65*(4), 262–278. https://doi.org/10.1037/a0019290

Vallacher, R. R., Michaels, J. L., Wiese, S., Strawinska, U., & Nowak, A. (2013). Mental dynamism and its constraints: Finding patterns in the stream of consciousness. In D. Cervone, M. W. Eysenck, M. Fajkowska, & T. Maruszewski (Eds.), *Personality dynamics: Embodiment, meaning construction, and the social world*. New York: Eliot Werner Publications.

Van de Vliert, E., & Euwema, M. C. (1994). Agreeableness and activeness as components of conflict behaviors. *Journal of Personality and Social Psychology, 66*(4), 674–687. https://doi.org/10.1037/0022-3514.66.4.674

Wall Jr., J. A., & Callister, R. R. (1995). Conflict and its management. *Academy of Management Journal, 21*(3), 515–558 https://doi.org/10.1177/014920639502100306

Wegner, D. M., Schneider, D. J., Carter, S. R., & White, T. L. (1987). Paradoxical effects of thought suppression. *Attitudes and Social Cognition, 53*, 5–13 https://doi.org/10.1037/0022-3514.53.1.5

Wegner, D. M., Schneider, D. J., Knutson, B., & McMahon, S. R. (1991). Polluting the stream of consciousness: The effect of thought suppression on the mind's environment. *Cognitive Therapy and Research, 15*, 141–152 https://doi.org/10.1007/BF01173204

Weider-Hatfield, D., & Hatfield, J. D. (1995). Relationships among conflict management styles, levels of conflict, and reactions to work. *Journal of Social Psychology, 135*(6), 687–698 https://doi.org/10.1080/00224545.1995.9713972

Weisbuch, G. (1992). *Complex systems dynamics*. Redwood City: Addison-Wesley.

Chapter 8
Modeling the Dynamics of Sustainable Peace

Larry S. Liebovitch, Peter T. Coleman, David Futran, Devin Lee, Tamar Lichter, Nicholas Burgess, Daniel Maksumov, and Celine C. Ripla

8.1 Sustaining Peace Is a Lot more than Just Ending War

Peace is not just the absence of war. For decades scholars in conflict resolution have studied the pathologies of war, violence, aggression, and conflict (Deutsch et al. 2014; Deutsch 1977, 2002; Kriesberg 2007; Pruitt et al. 2004). Peace has been studied only in the context of those processes. Very little is known about the fundamental conditions needed to sustain peace. We are therefore studying the complex, multidimensional, and dynamical processes that are needed to sustain peace.

L. S. Liebovitch (✉)
Departments of Physics and Psychology, Queens College, City University of New York, Flushing, NY, USA

Physics Program, The Graduate Center, City University of New York, New York, NY, USA

Adjunct Senior Research Scholar, Advanced Consortium for Cooperation, Conflict, and Complexity (AC4), The Earth Institute at Columbia University, New York, NY, USA
e-mail: larry.liebovitch@qc.cuny.edu

P. T. Coleman
Morton Deutsch International Center for Cooperation and Conflict Resolution (MD-ICCCR), Institute for Psychological Science and Practice (IPSP), Teachers College, Columbia University, New York, NY, USA

Advanced Consortium for Cooperation, Conflict and Complexity (AC4), The Earth Institute at Columbia University, New York, NY, USA

D. Futran · D. Lee · T. Lichter · D. Maksumov
Queens College, City University of New York, New York, NY, USA

N. Burgess
Queens College, City University of New York, New York, NY, USA

Columbia University, New York, NY, USA

C.C. Ripla
Northeastern University, College of Engineering, Boston, MA, USA

U. Strawinska-Zanko, L. S. Liebovitch (eds.), *Mathematical Modeling of Social Relationships*, Computational Social Sciences,
https://doi.org/10.1007/978-3-319-76765-9_8

8.2 The Sustaining Peace Mapping Project

In 2014, an interdisciplinary team of scientists, academics, policy-makers, and practitioners convened by the Advanced Consortium on Cooperation, Conflict, and Complexity (AC4) at Columbia University launched a multi-year initiative aimed to provide a more comprehensive and fundamental understanding of sustainable peace. The goals of this project are to:

- Use the scientific evidence from a wide range of disciplines to identify factors that influence sustainable peace
- Create a shared understanding of the relationships between the main factors influencing sustainable peace and their relative importance
- Build on this evidence to create an interactive causal loop diagram to identify effective interventions, measurable goals, and empirically and locally informed indicators for tracking trends in sustainable peace

The core team consists of researchers and practitioners with a broad range of expertise: Peter T. Coleman (social psychology), Joshua Fisher (geography and environmental science), Beth Fisher-Yoshida (communications), Douglas P. Fry (anthropology), Larry S. Liebovitch (physics and psychology), Philippe Vandenbroeck (philosophy), Danny Burns (international development), Kristen Rucki (international education), and Jaclyn Donahue (international development). The core group received extensive additional input from 72 responses on a survey from subject matter experts in a wide range of scientific fields (including neuroscience, evolutionary biology, political science, environmental policy, philosophy) and one small and one larger workshop that included participants from 9 universities in the United States, United Kingdom, and Turkey and representatives from the United Nations, Environmental Law Institute, the United States Institute of Peace, the Inter-American Development Bank, The Omidyar Group, and Bloomberg LP.

A central and evolving product of this work has been to identify the factors needed for sustainable peace in the world and how they influence each other. This is being represented in a visualization called a causal loop diagram. Fig. 8.1 is the current version of the causal loop diagram with the peace factors and their positive (+) and negative (−) influences on each other. The central peace factors PIR and NIR are the positive and negative intergroup reciprocities.

8.3 Mathematical Model of the Causal Loop Diagram of Sustainable Peace

The causal loop diagram is a qualitative description of how these peace factors influence each other. The central goal of the part of the project described here was to develop a rigorous mathematical model of the interactions between these peace factors and use it to determine the dynamics of this system.

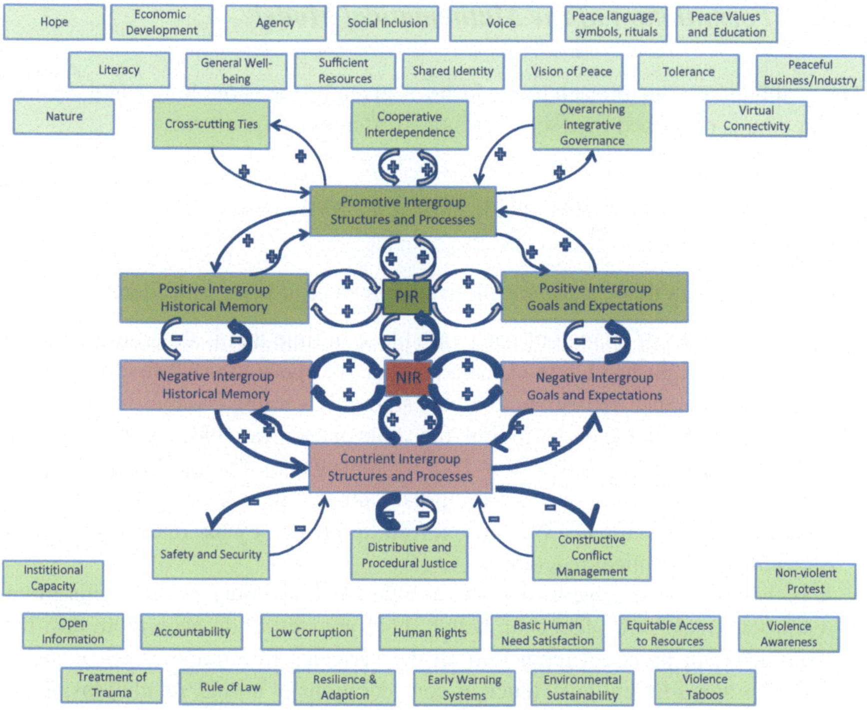

Fig. 8.1 The June 3, 2016, causal loop diagram of sustainable peace from AC4

8.3.1 The Value Added of a Mathematical Model

Transforming the qualitative causal loop diagram into a rigorous mathematical model provides valuable new insights for this project. A mathematical model may be able to:

- Reveal properties about a system that may be difficult to discern in a qualitative causal loop model
- Determine how the quantitative values of the peace factors depend on each other and evolve in time to understand how these peace factors operationally function as a system
- Make quantitative predictions on how the values of the peace factors would change in response to different interventions in the system, such as changing the values of some of the peace factors or the strengths of the influence between them
- Be the input into a graphic display of the quantitative values of the peace factors and the strengths of the connections between them so that researchers can clearly see and explore the effects of changes in the model

8.3.2 Formulation of the Mathematical Model

The quantitative value of each peace factor is given by the variable x_i, and its evolution in time is described by a set of ordinary differential equations:

$$\frac{dx_i}{dt} = -|m_i|x_i + b_i + \sum_{j=1}^{n} c_{ij} \tanh(x_j). \tag{8.1}$$

where each term in this equation is motivated by the following considerations:

- dx_i/dt is the rate of change of each variable x_i in time that is determined by the terms on the right-hand side of the equation. We chose to use a first derivative, rather than a higher derivative, to model smoothly varying changes in time.
- $-|m_i|x_i$ is a proportional decay to limit the value of each variable x_i proportionately to its current value, that is, if it is 100, we reduce it to 90, and if it is 1000, we reduce it to 900. We do this to prevent the values of the variables from possibly increasing without bound. We chose a form motivated by the rate of decay of a molecule in a chemical reaction.
- b_i is the self-reinforcement of each variable (such as doing positive intergroup reciprocity makes you feel even better about yourself to do even more positive reciprocity) or an equal input into all the variables from another set of meta-variables at a different level (such as considerations of norms, regulations, institutions, and constraints). We chose a constant value that best represents self-reinforcement for each variable or a constant input from the outside to all the variables.
- c_{ij} is the strength of the influence from variables j to i, that is, from x_j to x_i. We use a hyperbolic tangent function tanh(x) so that low values of each variable will have a proportionate influence on the other variables, but that influence reaches a threshold of maximum influence when the value of each variable is very high, as illustrated in Fig. 8.2. We chose this functional form as it has been useful in other computational structures, such as the connections between nodes in artificial neural networks.

We have also used similar equations in other models of 2 people interacting and 2000 people interacting (Liebovitch et al. 2008, 2011; Peluso et al. 2012; Fernandez-Rosales et al. 2015) so that we have considerable experience and understanding of both the analytical and numerical behavior of such equations.

8.3.3 Parameters of the Mathematical Model

We choose values for the parameters based on previously published studies and our own experience. For parameters where data is not available, we explore the properties of the model as those parameters are varied over a wide range of values.

Fig. 8.2 The influence of variable x_j on variable x_i

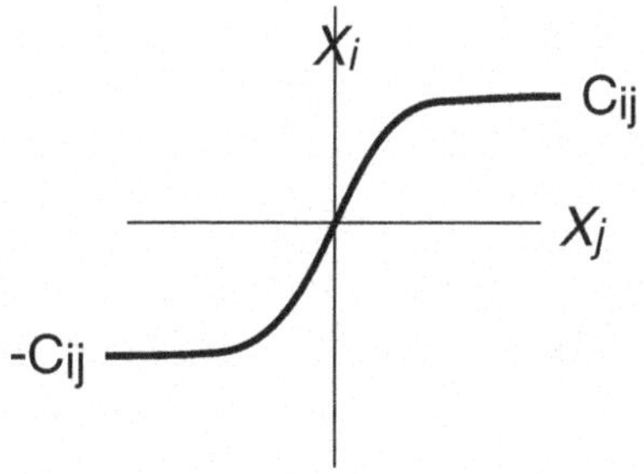

- c_{ij} is the strength of the influence from variables *j* to *i*. This is the most important parameter in transforming the qualitative causal loop diagram into a rigorous mathematical model. In a causal loop diagram, there is no quantitative value for the influence between the variables. In a rigorous mathematical model, the quantitative connection strengths between the variables determine if there is a single or multiple attractors (values of the variables at long times), the type of dynamics of the system (monotonic or oscillatory), and the sensitivity of the dependence on the initial values of the variables. These values will be estimated from published studies, as was done in the preliminary analysis described below.
- m_i is the time constant of exponential decay, which is the degree of memory of the system, and was also called the inertia to change by Gottman et al. (2005). What is most important here is the relative value of this parameter among the different variables. Gottman et al. (2005) found that negative memories have both a stronger influence, and their effects last longer than positive memories. For this reason we set $m_i = -0.2$ for the variables that represent such memories, such as negative historical memory, while we set $m_i = -0.9$ for all the other variables.
- b_i is the self-reinforcement of each variable or an equal input into all the variables from another set of meta-variables at a different level. As data for these parameters is not available, we will explore the effects on the system of a wide range of these values.
- Initial conditions are the initial values given to the variables at the beginning of the computation. We also vary these over a wide range to determine how the dynamics of the system depends on them.

8.3.4 *Solution of the Mathematical Model*

The equations were integrated numerically to determine the dynamics, the existence of steady states, and their dependence on initial conditions. We have experience in using many different numerical integration methods (fourth-order Runge-Kutta, predictor-corrector methods, and different finite-difference schemes). Here, because of its simplicity and stability, we chose to use the Euler integration with a suitable small $\triangle t$ step size:

$$x_i(t+\Delta t) = x_i(t) + \Delta t\left\{-|m_i|x_i(t) + b_i + \sum_{j=1}^{n} c_{ij}\tanh\left(x_j(t)\right)\right\} \tag{8.2}$$

We run the calculation from different initial conditions with different parameters to identify which variables play the most important roles in the dynamical behavior of the system and its long-term steady states (attractors).

8.4 Initial Results from a Mathematical Model of one Segment of the Causal Loop Diagram

We now present an example to demonstrate what additional information can be gained by using such a mathematical model. We constructed a mathematical model of the "core engine," a small but central component, from the April 4, 2016, version of the sustainable peace map.

8.4.1 Parameters of this Initial Mathematical Model

The variables in this model are:

Positive variables:

#1 – Positive historical intergroup memory
#3 – Positive goals and expectations
#5 – Positive intergroup reciprocity

Negative variables:

#2 – Negative historical intergroup memory
#4 – Negative goals and expectations
#5 – Negative intergroup reciprocity

The causal loop diagram is presented in Fig. 8.3:

Next, we determined the qualitative strengths of the connections between the variables. This was done by an extensive literature review to assign strengths of connections between variables as ordinal variables that were none, low, medium, medium-high, and high, as shown in Fig. 8.4. From our previous experience (Liebovitch et al. 2008), we know that given the values of the parameters m_i and b_i, the values of c_{ij} then define different regimes (bifurcations). From that information we can assign representative numerical values to the connection strengths c_{ij} from the ordinal values, as shown in Fig. 8.5.

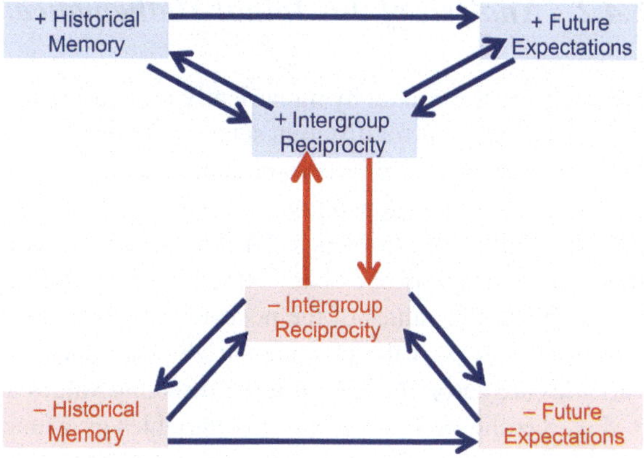

Fig. 8.3 Segment of the April 4, 2016, causal loop diagram

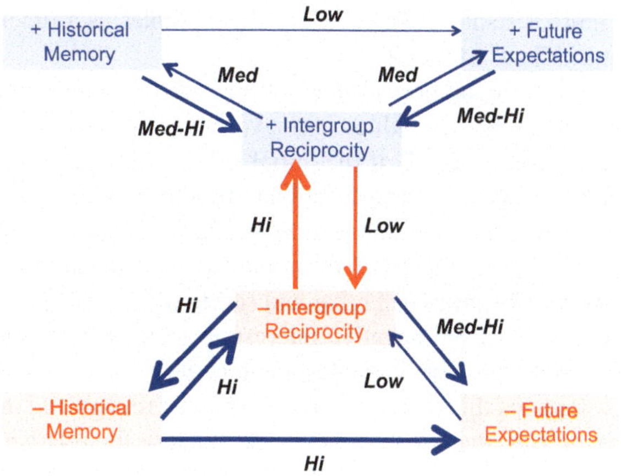

Fig. 8.4 Segment of the April 4, 2016, causal loop diagram with relative ordinal connection strengths

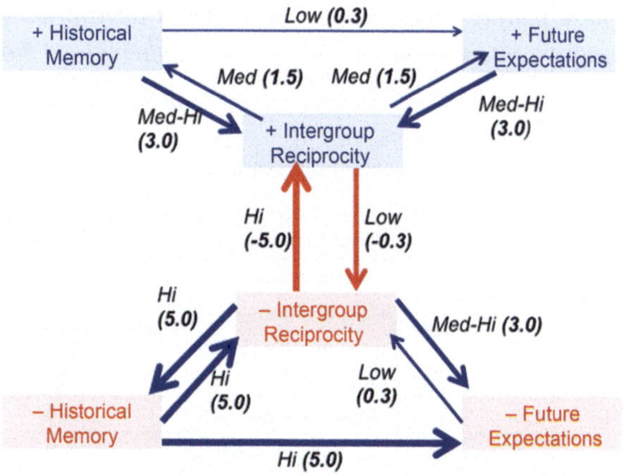

Fig. 8.5 Segment of the April 4, 2016, causal loop diagram with numerical connection strengths

8.4.2 Analysis of this Initial Mathematical Model

We analyzed this model by numerically integrating the equations from a wide range of initial conditions for different sets of parameters. The results of this mathematical model led us to several important conclusions.

Over long times, depending on their initial values, the values of these variables evolve to only one of two possible sets of values, which are called the attractors of this system. There is one "bad" attractor where the values of the negative variables are large and the values of the positive variables are zero and one "good" attractor where the values of the positive variables are large and the values of the negative variables are zero. These long-term attracting values are shown in Table 8.1.

Fig. 8.6 shows the values of the variables as a function of time approaching the "good" attractor. It also shows the strengths of the relative values of each variable at the attractor, as indicated by the size of the circles for each variable, as well as the relative strengths of the connections between the variables defined by the c_{ij} matrix. Figure 8.7 shows the values of the variables as a function of time approaching and at the "bad" attractor.

From the numerical integrations starting with many different initial conditions, we found that the values of the variables *almost always* end up in the bad attractor, at long times. (Only if the initial values of the negative variables are less than 0.01 does this system escape the bad attractor.) Why does this system always go to the bad attractor? The connection strengths of the negative variables are all stronger than those of the positive variables. That means that the negative variables will always reinforce each other and rise to higher values, and then the strongly negative connection from negative intergroup reciprocity to positive intergroup reciprocity forces the positive variables down to zero.

Perhaps this feature should have been recognized in the original causal loop diagram. But, that could not be done because the relative strengths of the connections between the variables are not defined in the original causal loop diagram. Perhaps this feature should have been recognized once the ordinal values of the connection strengths were determined. It was still far from obvious that would be the case. *These findings demonstrate the value of transforming the qualitative causal loop diagram into a quantitative rigorous mathematical model and then using the math-*

Table 8.1 Long-term values of the variables

Variable #	Variable name	Bad attractor	Good attractor
1	+ memory	0	1.7
3	+ expectations	0	2.0
5	+ reciprocity (PIR)	0	6.3
2	- memory	24.9	0
4	- expectations	8.9	0
6	- reciprocity (NIR)	5.9	0

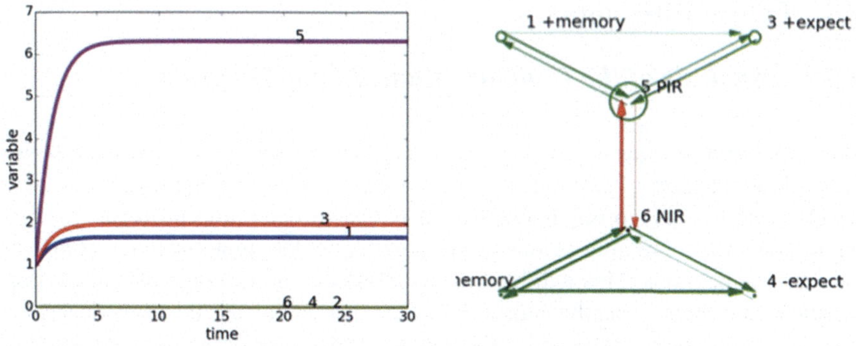

Fig. 8.6 Values of the variables approaching and at the "good" attractor

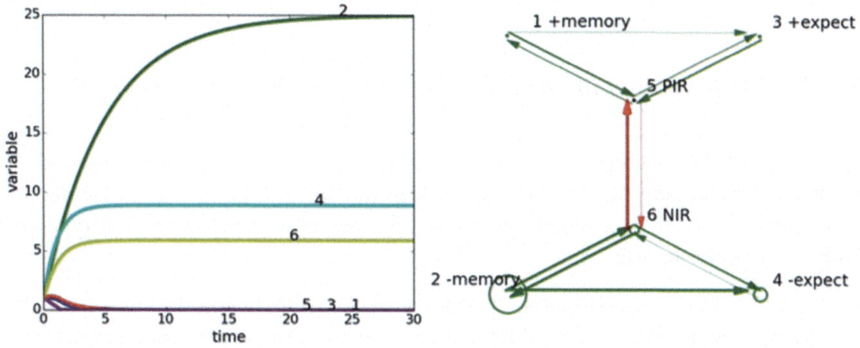

Fig. 8.7 Values of the variables approaching and at the "bad" attractor

ematical and computational methods to analyze the system to discover new and important information in the sustainable peace map.

We don't believe that it can always be bad. Thus, this analysis tells us that there is an important missing piece or pieces in this model. Additional numerical integrations with different parameter values showed that this system will go to a good attractor if:

- The individual variables reinforce themselves, such as doing good positive intergroup reciprocity makes you feel even better about yourself to do even better positive intergroup reciprocity.
- There is an equal input into all the variables from another set of meta-variables at a different level that represent considerations of norms, regulations, institutions, and constraints.

8.5 Future Directions

8.5.1 *Analysis of the Complete Causal Loop Diagram*

Our initial analysis was based only on a very small subset of the very much larger causal loop diagram of sustainable peace. A central goal of this project is to now develop and use a rigorous mathematical model to determine properties of this entire system that are not obvious or not possible to determine from the complete causal loop diagrams. This will be done by transforming the complete causal loop diagrams into rigorous mathematical models so that we can use the numerical solutions of the equations to determine the dynamical properties of system, the number of attractors, the initial conditions, the parameter values that lead to each attractor, the stable or unstable dynamics of the evolution of the values of the variables in time, and the sensitivity of the system to its respective parameters and variables.

8.5.2 *Graphic Display and Interactive Graphic Interface*

The utility of the mathematical model depends on the clarity and usefulness of its human-computer interface so that people can understand the results of the model and interact with it in a meaningful way. We are now developing two types of human-computer interfaces:

- A graphic display to illustrate dynamics of the model; how the values of the variables evolve over time, which can be viewed as both still frame graphs; and a time lapse animation where the value of each variable is indicated by a box whose size and/or color changes as the variable evolves in time
- An interactive point-and-click graphic display so that parameters and variables can be changed while the numerical integration is in progress, in a user-friendly way, by someone without computer programing expertise to enable practitioners and policy-makers to use the model to explore the effects of different possible interventions.

8.5.3 *Data Science to Measure the Variables*

We are now starting to use modern data science techniques to collect data from databases and social media to determine the quantitative values of the peace variables and to test the validity of the values predicted by the mathematical model. For example, the important variables of the strength of the positive and the negative historical memories can be measured from Facebook, Twitter, and trending Google searches for words that identify "past" or "future," and then a computer program does a “sentiment analysis” by evaluating the net emotional positivity or negativity of the words in those posts. In our preliminary studies so far:

- We have written scripts in Python and R that “scrape” data from Twitter feeds from specific individuals or within specific geographic areas. We have also used databases of the emotional content of specific words to identify the net positive or negative emotional content in text (Guzmán-Vargas et al. 2015), which will be applied to these tweets.
- Since many of the variables in the causal loop diagram are intergroup variables, we are also exploring ways to identify group membership. For example, to test our programs to define a membership group, we have used Force Atlas 2 and Gephi (Bastian et al. 2009) to construct the network of all the hashtags on Twitter that are linked to #BlackLivesMatter.

8.6 Other Applications of this Mathematical Model

Causal loop diagrams have been a valuable tool to analyze and understand the system properties and dynamics of complex systems. They have been used to analyze a wide variety of systems including political systems, the causes of genocide, the consequences of teenage pregnancy, and the choices in tackling obesity (Burns 2007; Foresight 2016; Ricigliano et al. 2016). The mathematical framework presented here has been helpful in providing new information about the causal loop diagram of sustainable peace. A similar mathematical approach can also have considerable value in analyzing the causal loop diagrams of a wide variety of other systems.

In a complex causal loop diagram, with many positive and negative feedback loops, it can be challenging to trace out the system-wide effects of changing the value of one variable, or a set of variables, or the feedback loops between the variables. The mathematical model developed here makes it possible to unambiguously determine:

- The long-term steady-state values of all the variables in the system
- The dynamics of how all the variables evolve in time, whether they monotonically approach final values or fluctuate periodically or chaotically around sets of values
- The response of the entire system, that is, the values of all of the variables, to changes in the initial values of the variables or the feedback loops between the variables
- The response of the entire system to an intervention that uses inputs from outside the system to hold a given variable, or a set of variables, at a fixed value

Applying this mathematical model to other systems requires:

- Operational definitions of the variables so that they can be measured

- Estimates or measurements of the quantitative values of the strengths of the connections between the variables
- Sufficient computational resources to numerically integrate the equations and determine their dependence on a wide range of initial conditions and system parameters
- A suitably intuitive graphic display of the results of those computations and an interactive human-computer interface so that the analysis can be appreciated and policy interventions explored by non-technologically sophisticated users

The mathematical formulation we have presented here can be viewed as a general computational structure, using (hyperbolic tangent) transfer functions, similar to those used in artificial neural networks, to link together and therefore compute the effects of the variables acting on each other. In that sense, this computational structure shares properties with artificial neural networks and machine learning systems that transform input data (the initial values of the variables and the strengths of the connections between them) into output data (the final values of the variables and their time course in reaching them) and so it should have broad applicability to a wide range of applications.

References

Bastian, M., Heymann, S., & Jacomy, M. (2009). *Gephi: An open source software for exploring and manipulating networks*. International AAAI Conference on Weblogs and Social Media. https://gephi.org/users/publications/

Burns, D. (2007). *Systemic action research: A strategy for whole system change*. Bristol: Policy Press.

Deutsch, M. (1977). *The resolution of conflict: Constructive and destructive processes*. New Haven/London: Yale University Press.

Deutsch, M. (2002). Social psychology's contributions to the study of conflict resolution. *Negotiation Journal, 18*(4), 307–320. https://doi.org/10.1111/j.1571-9979.2002.tb00263.x

Deutsch, M., Coleman, P. T., & Marcus, E. C. (Eds.). (2014). *The handbook of conflict resolution: Theory and practice*. Wiley.

Fernandez-Rosales, I. Y., Liebovitch, L. S., & Guzman-Vargas, L. (2015). The dynamic consequences of cooperation and competition in small-world networks. *PLoS One, 10*(4), e0126234. https://doi.org/10.1371/journal.pone.0126234.

Foresight: Obesity Systems Map. (2016). https://www.gov.uk/government/uploads/system/uploads/attachment_data/file/296290/obesity-map-full-hi-res.pdf. Accessed 19 Aug 2016.

Gottman, J. M., Murray, J. D., Swanson, C. C., Tyson, R., & Swanson, K. R. (2005). *The mathematics of marriage: Dynamic nonlinear models*. Cambridge, MA: MIT Press.

Guzmán-Vargas, L., Obregón-Quintana, B., Aguilar-Velázquez, D., Hernández-Pérez, R., & Liebovitch, L. S. (2015). Word-length correlations and memory in large texts: A visibility network analysis. *Entropy, 17*(11), 7798–7810. https://doi.org/10.3390/e17117798.

Kriesberg, L. (2007). *Constructive conflicts: From escalation to resolution*. Lanham: Rowman & Littlefield.

Liebovitch, L. S., Naudot, V., Vallacher, R., Nowak, A., Bui-Wrzosinska, L., & Coleman, P. (2008). Dynamics of two-actor cooperation-competition conflict models. *Physica A, 387*, 6360–6378. https://doi.org/10.1016/j.physa.2008.07.020.

Liebovitch, L. S., Peluso, P. R., Norman, M. D., Su, J., & Gottman, J. M. (2011). Mathematical model of the dynamics of psychotherapy. *Cognitive Neurodynamics, 5*(3), 265–275. https://doi.org/10.1007/s11571-011-9157-x.

Peluso, P. R., Liebovitch, L. S., Gottman, J. M., Norman, M. D., & Su, J. (2012). A mathematical model of psychotherapy: An investigation using dynamic non-linear equations to model the therapeutic relationship. *Psychotherapy Research, 22*(1), 40–55. https://doi.org/10.1080/10503307.2011.622314.

Pruitt, D., Rubin, J. Z., & Kim, S. H. (2004). *Social conflict: Escalation, stalemate, and settlement.* New York: McGraw-Hill.

Ricigliano, R., Wicken, S., & Glaisyer, T. (2016). *Case studies-systems thinking in practice.* International Conference for Conflict Management, June 28, 2016, New York.

Chapter 9
Capital in the First Century: The Evolution of Inequality in Ancient Maya Society

Urszula Strawinska-Zanko, Larry S. Liebovitch, April Watson, and Clifford T. Brown

9.1 Introduction

> The distribution of wealth is one of today's most widely discussed and controversial issues. But what do we really know about its evolution over the long term? (Piketty 2014:1).

These are the opening lines of Thomas Piketty's influential recent book *Capital in the Twenty-First Century* (2014). Truer words could not be written, but we have a strikingly different interpretation of the phrase "long term." Whereas Piketty looks back as much as 300 years, we start our investigation about 2000 years ago to look for the origins of inequality in an early civilization, and we encourage others to use the techniques we deploy here to push the research yet further back in time, to 4000 or 5000 (or more) years ago in other parts of the world. For we believe that critical facts about inequality, such as its origin, its evolution, and its consequences, cannot

U. Strawinska-Zanko (✉)
Department of Conflict Resolution Studies, College of Arts, Humanities, and Social Sciences, Nova Southeastern University, Fort Lauderdale, FL, USA
e-mail: uzanko@nova.edu

L. S. Liebovitch
Departments of Physics and Psychology, Queens College, City University of New York, Flushing, NY, USA

Physics Program, The Graduate Center, City University of New York, New York, NY, USA

Adjunct Senior Research Scholar, Advanced Consortium for Cooperation, Conflict, and Complexity (AC4), The Earth Institute at Columbia University, New York, NY, USA

A. Watson
Lynn University, Boca Raton, FL, USA

C. T. Brown
Florida Atlantic University, Boca Raton, FL, USA

U. Strawinska-Zanko, L. S. Liebovitch (eds.), *Mathematical Modeling of Social Relationships*, Computational Social Sciences,
https://doi.org/10.1007/978-3-319-76765-9_9

be understood without the *very* long-term perspective uniquely provided by archaeology. We will argue here that a special degree of inequality is characteristic of state societies. Archaeologists already know that ancient states exhibit dramatic inequality. Anyone can see that by merely looking at ancient royal palaces and tombs. We, however, develop a novel and more specific argument in which we show that one particularly extreme distribution of wealth, known as a power law, is characteristic of the fully developed state.

Inequality has long been a central theme in all the social sciences, including archaeology. The degree and kind of inequality distinguishes rank societies from egalitarian ones (Ames 2007), chiefdoms from tribes, and states societies from all the others. Inequality can take various forms. Material wealth can be differentially distributed, but so can political power, social status, prestige, religious influence, and other factors that contribute to or determine affluence and reproductive success (Bowles et al. 2010). We are investigating the emergence of economic inequality because it is a dominant but poorly understood feature of the development of complex societies, particularly those politically centralized, urbanized polities that we call states or, more broadly, civilizations. Understanding the form that inequality took is crucial to inferring how it developed, and both are essential to disentangling the interwoven processes involved in the evolution of social complexity. In this study, therefore, we are trying to determine not only the overall extent of inequality but particularly whether the ancient Maya exhibited the modern, Pareto, form of the wealth distribution; if so, what its parameters were and when it emerged in their society.

We describe the distributions of wealth at four ancient lowland Maya archaeological sites, Komchén, Palenque, Sayil, and Mayapán (Fig. 9.1), and we discuss how the distributions changed through time. We used house sizes as a measure of wealth, an imperfect proxy but one that is difficult to improve upon without major investments in new fieldwork. We calculated the Gini coefficient for the datasets as one indicator of the equality of the distribution of wealth, but we focused on ascertaining whether the distributions corresponded to Pareto (power-law) distributions, such as we find in modern industrial societies, because determining the functional form of the distributions can provide insights into the dynamic processes that gave rise to them. We have previously reported the analysis of the datasets from three of these sites, Palenque, Sayil, and Mayapán (Brown et al. 2012). In this article, we reanalyze those datasets using maximum likelihood estimation, and we add a fourth site, Komchén. Komchén is particularly important because it extends our analysis back into the Late Preclassic period, to a time when state-level, urban society may have begun to arise in other parts of the Maya area but before it developed in northern Yucatán (Anderson 2005, 2010). The data from Komchén therefore allows us to observe the evolution of inequality across the critical transition from prestate to state-level society. It also helps us to address the much debated question about the degree of economic inequality in prestate societies such as chiefdoms.

Modern industrial societies display gross economic inequality. Distributions of wealth are not the result of simple random variation in wealth, which would yield a "bell curve" or normal distribution representing a society with a few poor, a few

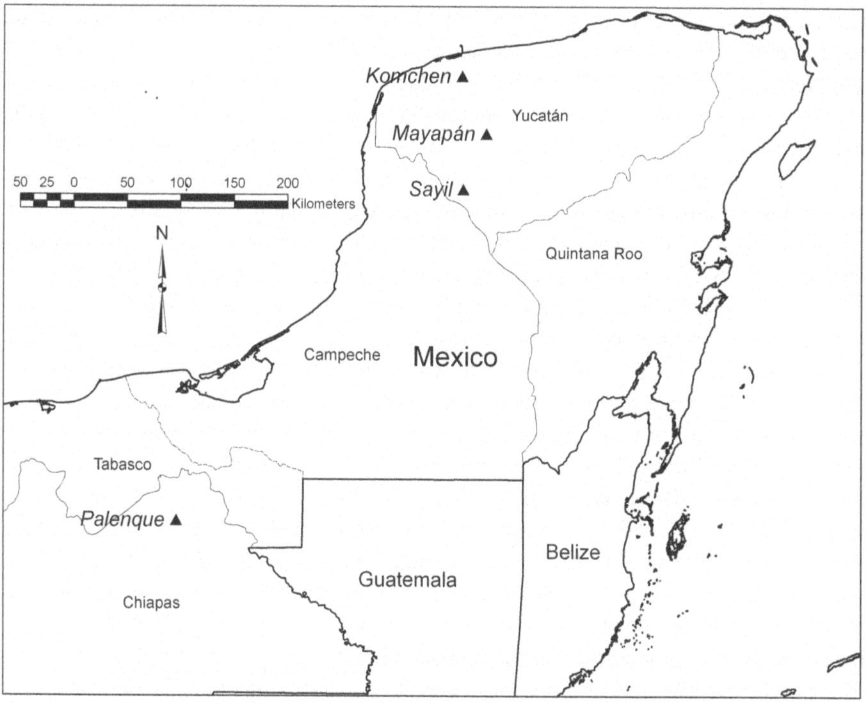

Fig. 9.1 Map of the Maya lowlands showing the locations of the four sites analyzed

rich, and a large middle class. The actual observed pattern is quite different, with a relatively few extraordinarily rich and ever-increasing numbers of people in ever-lower income classes. So pervasive is this pattern that it may appear inevitable, part of the natural human condition, but that is not so.

Egalitarianism has been amply documented in modern, historic, and ancient egalitarian societies. Thus, we know that humans are not ineluctably condemned to live in conditions of gross inequity. Egalitarian societies are, and were, typically small-scale kin-based societies that lack an overweening political structure. They practice collective ownership of the resources or the means of production; they possess deeply ingrained norms of hospitality, sharing, prestation, and generalized reciprocity that create equality in the face of inevitable variation; and they deploy social sanctions and controls—gossip, ridicule, and sometimes even violence—to maintain it (Lee 1990; Trigger 1990; Wilk 1983). Acute inequality is not, therefore, the only possible configuration for human society. Egalitarianism preceded marked inequality and even persists today, illustrating the durability of the associated social logic.

Social inequality emerged from egalitarian societies in the form of rank societies, stratified societies, chiefdoms, or what are generically called "middle-range societies." Many chiefdoms seem to have conceptualized two or three hierarchically arrayed social classes but these classified people into social ranks that did not represent economic classes (Service 1971). Despite a large body of knowledge about these societies (Ames 2007; Drennan and Peterson 2006; Earle 1991, 1997; Feinman and Neitzel 1984; Marcus 2008), it is still unclear whether they were also characterized by economic inequality or whether differentiation in wealth only emerged with the rise of state societies (Drennan and Peterson 2012; Earle 1997; Yoffee 2005). Through our analyses of Maya sites ranging from Late Preclassic to Late Postclassic, we tried to establish the timing of the emergence of economic inequality.

One of the core characteristics of the early states was extreme economic inequality (Smith 1987; Yoffee 2005) coexisting with social stratification in which social classes were often but not always formally and juridically defined and in most cases hereditary. Given limited social mobility, social classes were at least as important at determining who became rich as wealth was in determining who belonged to which social class. Because social status was so often hereditary, social status and wealth were probably not always perfectly correlated, yet the two were closely aligned (Trigger 2003; Yoffee 2005). Ethnohistorical research reveals that at least in the Postclassic period (ca. A.D. 1000–1542), the Maya almost certainly had three hereditary classes, nobility, commoners, and slaves (Marcus 1992, 1993, 2004; Roys 1957; Thompson 1999), and they might have recognized a middle class as well, although the documentary evidence for it is exiguous (Roys 1957). Whether these social classes existed earlier, during the Classic period (A.D. 250–900) or before, is partly conjectural, but the hieroglyphic inscriptions do employ a number of terms for royalty and nobility that suggest gradations in status. Despite the historical evidence for institutionalized social classes, the archaeological evidence for the distribution of wealth, for example, in household assemblages of artifacts at Mayapán, implies a continuous distribution rather than one punctuated by discontinuities representing discrete classes (Masson and Pereza Lope 2005). A similarly broad and continuous distribution has been found at Coixtlahuaca, in Central Mexico (Steere and Kowalewski 2012:20), a contemporaneous site loosely culturally affiliated with Mayapán.

The extremely unequal distribution of wealth found in modern industrial societies was first described mathematically by Italian sociologist Vilfredo Pareto (1897, 1971), who studied the distribution of income. His formulation corresponds to a generic type of distribution that today we call a power law, or sometimes, eponymously, a Pareto distribution. In this distribution, an ever-smaller fraction of the population has an ever-larger amount of the wealth. Quantitatively, the fractional increase in wealth is proportional to the fractional decrease in the number of people with that wealth.

Since at least the 1950s, econometricians have proposed a number of models to explain the presence of the power-law distribution (e.g., Champernowne 1953). As early as 1960, Benoit Mandelbrot, whose work pioneered the fractal theory of scaling in finance, explored the relationship between the Pareto curve and the stable distributions studied by Paul Lévy. In recent years, econophysicists have explained Pareto distributions of wealth primarily in terms of exchange processes (e.g., Bagrow et al. 2008; Ispolatov et al. 1998), although factors such as inheritance are clearly relevant too (Smith et al. 2010; Stiglitz 1969). One of the best known and most influential dynamical models is Bouchaud and Mézard's (2000) model of wealth condensation, which predicts that exchange tends to reduce inequality. An agent-based archaeological version of this model has been elaborated by (Bentley et al. 2005), who found that trade can produce extreme inequalities in wealth, taking the form of power-law distributions, when exchange is based on "best price" rather than on notions simulating reciprocity.

A different explanation for the emergence of Pareto distributions of wealth has been proposed by Manus Midlarsky in his book *The Evolution of Inequality* (Midlarsky 1999). He attributes the emergence of a power-law distribution of wealth to land scarcity in agricultural societies. His argument is more geometric than algebraic in conception and focuses on the distribution of landholdings, that is, spatial or cadastral patterns of land tenure, which were the primary and ultimate source of wealth in early civilizations (Trigger 2003). Midlarsky observes that statistically fractal patterns of landholdings will yield a Pareto distribution of wealth, and the fractal patterns are easily generated by simple and plausible principles. Specifically, he argues that in its initial stages, an agrarian society will enjoy an excess of arable land or other resources. In the absence of scarcity or competition, the resulting distribution of wealth will be affected primarily by natural variation in both the quality of the resource and the human population. Significant but not extreme inequality should develop under these conditions, yielding an approximately exponential distribution of wealth. However, growing population would inevitably create scarcity in a finite, immovable, and vital resource such as land, eventually leading to a fractal pattern of land tenure, which in turn would create a Pareto distribution of wealth. A subsequent decrease in population could have an ameliorating effect on inequality (Midlarsky 1999). Midlarsky has found Pareto distributions of land ownership in modern countries that are characterized by extreme inequality in the distribution of wealth. We will explore the implications of these models for our case studies in the closing discussion.

In summary, we asked:

1. Did ancient Maya societies exhibit a Pareto distribution of wealth?
2. If so, can we identify when it emerged?
3. How did it change through time?

9.2 Methods

One can talk about inequality as a moral or social or political problem, and one can philosophize about it, as many do, in the abstract….But you can't actually study economic inequality without measuring it. (Galbraith 2012:7).

We used house sizes as an index of wealth (Abul-Magd 2002; Hayden 1997; Hutson et al. 2006; Kurjack 1974; McGuire 1983, 2001; Rathje 1983; Rathje and McGuire 1982; Smith 1987; Yoffee 2005). There are excellent maps of buildings at archaeological sites in the lowland Maya area that provide a sufficiently large sample of data for the analysis of wealth distributions, which would be difficult to achieve if a different index of wealth was used, such as mortuary furniture from burials or domestic artifacts from household excavations. Houses are often considered the strongest and most consistent expression of wealth levels in agrarian states (Hirth 1989; Smith 1987). Nonetheless, other variables can confound the equation between the distribution of house sizes and the distribution of wealth. For example, houses located in higher prestige areas may be more valuable than those elsewhere, house size may reflect family size more than wealth, and factors such as zoning regulations may affect house sizes. To justify the use of house size as an index of wealth in the context of ancient Maya urbanism, it is important to acknowledge the possible relationships between house sizes and other relevant social and cultural variables. We discuss intra-site location, zoning, postmarital residence, and population. Therefore, we first review the factors and evidence that support the use of house size as a measure of wealth.

It is possible that location had an influence on the value of houses, as it does in modern real estate. Although this may seem to be a projection of modern culture upon a radically different ancient one, it is nevertheless a legitimate question. For example, we might expect to see finer or more elaborate houses in more desirable locales, such as near the site center, along *sacbes* (elevated roadways), around marketplaces, or close to schools (called *calmecacs* by the Aztecs). Maya sites do exhibit a degree of concentric zoning, with larger, more elaborate houses (such as those with vaulted roofs) tending to occur near the civic-ceremonial center and smaller, simpler dwellings occupying the outer circles (e.g., Cobá (Folan 1983) and Dzibilchaltún (Kurjack 1974)). But the urban pattern is not simple or linear because large houses are found scattered throughout the sites—witness, for example, the huge Olvidado platform near the western edge of Palenque (Barnhart 2001)—and small houses do occur in and around ceremonial centers, such as Structures Q-15, Q-16 a and b, Q-63, Q-91-94, Q-222, Q-206, and many others, at Mayapán. In fact, overall site settlement patterns are characterized as much by "clumpiness" as by zonation. Sites are composed of patio groups (*plazuelas*) of domestic structures, probably occupied by extended families, which typically include multiple dwellings of varying size. Multiple patio groups then agglomerate into larger clusters containing a range of large and small houses. The clusters in turn combine to make neighborhoods or wards, that in turn clump together to form bigger units, such as quarters, that comprise the site itself. In other words, the settlement pattern is self-similar: the patio group is a microcosm and building block of the settlement clusters, which in turn form neighborhoods, which then constitute quarters, which comprise the site (Brown and Witschey 2003). This settlement pattern is traditionally interpreted as mirroring the hierarchical branching of the kinship and residence system. Thus, with the possible exception of the site epicenter, which might have been occupied by hereditary nobility at some sites, the Maya city does not exhibit obvious patterns of spatial segregation by wealth. It does not have slums like Whitechapel or Hell's Kitchen counterpoised against Belgravias or Park Avenues.

Maya settlement patterns are also usually spatially dispersed: with few exceptions, Maya sites have low urban population densities, such as those seen at Tikal,

Copán, Calakmul, Dzibilchaltún, and Cobá, all major Classic period sites. Because structures were not closely packed in space, room for expansion remained. To the extent that the wealthy did live in central locations, that tendency was probably the result of inherited social rank—nobility—rather than of market mechanisms. Indeed, Diego de Landa's description of Mayapán explicitly says that the hereditary lords lived in the center of the site (Tozzer 1941). Moreover, the concentric zoning may be in part due to the dynamics of settlement growth. The central nucleus of the site is often the oldest sector, and therefore dwellings grew progressively larger over centuries as they were successively refurbished and expanded. Thus, the archaeological evidence for market-driven housing values related to location is weak and ambiguous. The historical evidence for such processes is also weak. Many of the earliest Maya documents we possess (Barrera Vásquez et al. 1984; Documentos de Tabi n.d.; Roys 1939, 1943) concern land tenure, and they reveal bitter conflicts over territory among nearby communities. However, they overwhelmingly focus on agricultural lands and orchards rather than urban houses. Analyses of Colonial period Maya wills and testaments, which might reveal house values, are not much help either (e.g., (Restall 1995, 1997; Thompson 1999)). The majority of the documents come from the end of the period, the late 1700s and early 1800s. By that time, the Maya economy had transformed significantly under Colonial rule, and therefore its relationship to the pre-Hispanic Maya economy is uncertain. Moreover, though the wills provide values for house lots, they include wells, fruit trees, and buildings, but do not give values for structures themselves. In sum, the effect of location on house value is difficult to infer, but the existing evidence does not suggest a significant influence.

It is also possible that a set of legal prescriptions and/or proscriptions resembling modern zoning ordinances might have influenced settlement patterns by enforcing ethnic, industrial, caste, or class-based clustering. This kind of clustering is common but not universal in ancient and historic cities (York et al. 2011). As we have seen, at Mayapán the hereditary nobility lived in the center of the city, but their wealth depended on their noble status rather than the reverse. We do know that in central Mexico, there was a long history of urban barrios spatially organized along ethnic lines (such as the Oaxaca barrio at Teotihuacan) or along occupational lines, such as the *calpulli* of Aztec Tenochtitlán and Otumba that were organized into neighborhoods by their craft specialization (York et al. 2011). Despite long-term interaction between central Mexico and the Mayas, the cities of the Maya lowlands do not exhibit these same patterns with clarity. Archaeologists have searched with little success for ethnic or occupational wards at lowland Maya sites. Though we should not confuse the absence of evidence with evidence of absence, we cannot assert that much zoning existed. Moreover, what little we know of Maya law (Gubler 2000/2001; Tozzer 1941) is silent on the matter.

The size of houses has traditionally been thought to correlate with the number of inhabitants, and therefore it is often used as a measure of population. Raoul Naroll's original study (Naroll 1962) found a correlation between house floor area and number of inhabitants in a small ($n = 18$) cross-cultural ethnographic dataset. According to an allometric regression formula presented in his work, one inhabitant should be inferred for each 10 m^2 of floor area in a house. This study has been extensively used and cited in making prehistoric population estimates around the world, including in the lowland Maya area. Subsequent research (Brown 1987; Casselberry 1974; LeBlanc 1971; Wiessner 1974), however, has identified numerous problems with Naroll's study, from data coding errors to the use of an allometric rather than a linear model. These studies suggest that Naroll's original constant was incorrect, and they question the validity of attempts to calculate a single culturally universal scaling exponent that describes the relation between house size and population. Instead, it seems likely that different types of societies (such as mobile foragers, shifting horticulturalists, and subsistence agriculturalists) have such different residence forms that they should be considered separately rather than together. Cross-cultural ethnographic studies have shown that floor area displays associations with other social and cultural factors, such as postmarital residence patterns (Brown 1987; Divale 1977; Ember 1973) and wealth (Hayden and Cannon 1984; Wilk 1983). These investigations have shown that matrilocal societies have significantly larger houses than patrilocal ones and that house size can correlate with wealth. Robert Netting (1982) and Richard Wilk (1983) have shown, moreover, that, in agrarian and pastoralist societies, household size correlates with wealth, which may explain at least partially the house size-population effect observed by Naroll and others. Specifically, house size and population covary because they both depend upon wealth. In sum, one cannot simply assume in general that house size correlates well with number of inhabitants because size also relates strongly to other socioeconomic variables, such as social structure (i.e., residence patterns), mode of subsistence (i.e., foraging, pastoralism, horticulture, agriculture), and wealth. It is necessary, therefore, to consider cultures individually, as we do here.

The preponderance of ethnographic and ethnohistorical evidence points to a patrilocal postmarital residence pattern among the Maya (see Brown 1999 for an analysis). The Yucatec and other Maya groups at one time had a matri-patrilocal system whereby the newly married couple would live with the bride's family for some years, while her husband performed bride service (known as *haancab* in Yucatec) for her family (Wilk 1988). However, after the *haancab*, the couple resided permanently with the husband's family in a patrilineal compound. We need not, therefore, be concerned that variation in residence patterns is responsible for differences in house size within or among communities.

In the Maya case, Richard Wilk's ethnoarchaeological studies (Wilk 1983) indicate that house size correlates primarily with wealth and only secondarily with population. He noted this pattern even in a nominally egalitarian (or incipiently stratified) community that exhibited differences in wealth. However, he also observed a difference between the "egalitarian" community and more overtly stratified ones. In the former, house sizes varied in floor area, but all the houses were constructed in the same manner of the same materials, whereas in the latter case, the overt display of status and wealth differences led to variation in house types, their architecture, and construction materials.

Although we have been discussing floor area and house size, usually defined as the area covered by the roof of the dwelling, a potentially more robust measure of

Table 9.1 Characteristics of the four sites analyzed

Site	Location	Period	Occupation	Population
Komchén	Yucatán, Mexico	Late Preclassic	350–150 B.C.	2500–3000
Palenque	Chiapas, Mexico	Late classic	A.D. 600 to A.D. 800	4000–6000
Sayil	Yucatán, Mexico	Terminal classic	A.D. 800 to A.D. 1000	10,000–11,000
Mayapán	Yucatán, Mexico	Late Postclassic	A.D. 1200 to A.D. 1450	10,000–15,000

wealth has also been developed: the labor investment required to construct each house. It is obtained by employing local laborers and masons to build replicas of ancient Maya structures and documenting the costs (Abrams 1989, 1994; Erasmus 1965). Those data can then be used to infer construction costs for ancient buildings, as has been done at Copán, Honduras (Abrams 1989, 1994) and Sayil, Yucatán (Carmean 1990; 1991a, b). This labor investment approach, while excellent, is difficult to apply to structures that have not been excavated, and since archaeologists typically only excavate small numbers of buildings at a given site, the sample sizes produced by this method are inadequate for detailed distributional analyses. We have therefore relied on house sizes measured from archaeological maps or reported by investigators. In two cases for which data were available, we have also used house volumes, that is, the volume of construction in cubic meters of the residential structure including its substructure platform, if any. As others have done (e.g., Hutson et al. 2006), we interpret volume as an estimate of labor investment, albeit an inexact one.

We collected and analyzed data on house areas from four lowland Maya archaeological sites, Komchén, Palenque, Sayil, and Mayapán (Table 9.1). For Komchén and Palenque, we also have the volumes of structures, and we have analyzed those two datasets as well. We previously published a less elaborate analysis of the area data from Palenque, Sayil, and Mayapán (Brown et al. 2012). In the current paper, we report additional analyses using maximum likelihood estimation of those three datasets, as well as for the first time the analysis of the Komchén data, a critical site that allows us to project our analyses back into the Preclassic period, to a time preceding the emergence of state-level sociopolitical organization in that area (Anderson 2005, 2010). It is only, therefore, with the Komchén data that we are able to examine how the Maya wealth distribution behaved as civilization arose.

9.3 Sites

Komchén, Yucatán, México, is the largest known Preclassic (Formative) period site in the Northwest Karst Plain of the northern plains physiographic zone of the Maya lowlands (Dunning et al. 1998), an area with low annual rainfall and thin, poor soils, which limit its agricultural potential. These ecological deficits should have restricted human population density, but the region saw some of the densest human settlement in the ancient Maya world. In the 1980s, extensive mapping and excavation were carried out by E. Wyllys Andrews V and Norberto González Crespo with the goal of learning more about the origin and evolution of Maya civilization. The site is approximately circular with a radius of 800 m and an area of ~2 km^2 (Ringle 1985; Ringle and Andrews V 1990). A detailed map of 1 km^2 revealed 505 structures. The area mapped consisted of one broad rectangle that included the most of the larger buildings in and around the site center and also extended west to the periphery of settlement. Three narrower transects running north, south, and east from the site center were also mapped. The exceptionally thin soils virtually guarantee that all structures were identified. A larger proportion of the site core was mapped than of the peripheries. As the larger buildings are concentrated in the core, smaller structures are probably underrepresented in the sample.

Small excavations for the purpose of dating were made in a spatially stratified random sample of 152 of the mapped structures; another 10 structures of interest were partly excavated. Almost all of the excavated structures that produced an adequate sample of potsherds dated from the Middle to Late Preclassic periods, although several test pits did not yield any ceramics, while some others did not produce a sample of pottery sufficient to infer a date. The site was first occupied during the Ek phase, now believed to date from about 1000–700 B.C., at the beginning of the Middle Preclassic period (Andrews V and Bey III 2011). The peak of population occurred during the Late Preclassic Nabanche phase (ca. 350–150 B.C.), when the site was probably home to 2500–3000 people (Ringle and Andrews V 1990). The settlement declined rapidly in the next few centuries and was effectively abandoned by the beginning of the Early Classic period (A.D. 250). The site was partly reoccupied during the Late to Terminal Classic Copo phase (ca. A. D. 600–1000). About 15% of the structures sampled, mostly small ones, yielded only Late or Terminal Classic pottery, while some of the Preclassic platforms were reused as substructures for later buildings.

We chose Komchén largely because it is one of the few Late Preclassic sites in which the architecture was not subsequently buried underneath major Classic period constructions. The architectural remains are therefore visible to be mapped and studied. The extensive, precise map and the probabilistic sampling of the structures offered the best existing dataset for this early period. We took the structure area and volume data from Appendix 4 of William Ringle's dissertation on Komchén (Ringle 1985) and compared them to the structures as shown on the site map (Andrews V et al. 1980). We excluded three subsets of structures from the complete inventory: (1) 5 large and complex structures (21 J1, 24G1, 23F1, 21F1, 25O1) that did not appear to have been primarily residential (Ringle 1985; Ringle and Andrews V 1988), judging by their architectural configuration; in general, most of the other structures are thought to have been residential because of the presence of metates and other domestic remains (Ringle 1985; Ringle and Andrews V 1990); (2) 24 structures that dated exclusively to post-Formative periods (14R4, 6 M1, 13 J1, 15 M2, 15R1, 8 N1, 14R3, 23Z1, 18 M1, 15F2, 22P1, 16D2, 18H1, 27R2, 22G2, 18 K1, 27O1, 14F2, 27Q1, 18F1, 23P1, 14O1, 19 K1, 22Y1); and (3) the 153 structures measuring less than 20 m^2 in area. We think that structures so small were unlikely to be dwellings. The excavators considered structures less than 15 m^2 in area to be too small to be residential (Ringle and Andrews V 1990). We chose a slightly higher figure so that our four datasets would be consistent. We drew the line at 20 m^2 for the three post-Formative sites because that size has been proposed as part of the definition of the “minimum residential unit” in the Maya lowlands (Ashmore 1981). Moreover, even (Ringle and Andrews V 1990) expressed reservations about their 15 m^2 criterion, thinking that it might be slightly too conservative. Consistency among the four datasets enables meaningful and important comparisons among them, while using the 20 m^2 criterion does not significantly dilute our understanding of Komchén. The final number of houses in the sample analyzed from Komchén was 324.

The second site, Palenque, Chiapas, Mexico, is located on a narrow shelf or bench perched on a north-facing escarpment that overlooks the alluvial plains of Tabasco. We chose Palenque because of the excellent archaeological map made by Edwin Barnhart (Barnhart 2001). Although various earlier maps of the site had been made, they were incomplete. Barnhart surveyed the entire 2.2 km^2 extent of settlement, save for a far western extension (< 1 km^2) on private land to which he could not obtain access. It is possible that more settlement exists at the foot of the escarpment on the alluvial plains, but if it does, it has eluded the archaeologists searching for it; it may be deeply buried by alluvium and colluvium. Barnhart argues (Barnhart 2001) that the mapped area constitutes only the “core” of the site and that more settlement probably exists on other landforms, such as nearby hilltops, ridges, and the alluvial plain below. Archaeological survey of surrounding areas (Liendo Stuardo 2005) found abundant rural settlement, but no major centers nor any direct extensions of the Palenque urban zone. Barnhart discovered that Palenque had one of the highest settlement densities known for a Classic period site, thrice that of Sayil and a little higher than Komchén, though still much lower than Mayapán. Nevertheless, the total population was low because of the relatively small extent of the settlement. He estimated a peak population of about 4000–6000 people, a little larger than Komchén’s, but much lower than Sayil’s (ca. 8000–10,000) or Mayapán’s (ca. 10,000–15,000).

We extracted the dimensions of all the structures listed in Appendix A of Barnhart’s dissertation and then excluded those that appeared to be nonresidential judging by his descriptions and comments in the text and their appearance on the map. Because the structures at Palenque often lack obvious surface features, such as the walls, benches, and columns that are common at some sites (such as Mayapán), it was frequently difficult to distinguish residences from other types of buildings. The lack of surface features is probably partly attributable to the high rate of colluviation at the site, which has partially buried many of the structures. Again, we excluded structures smaller than 20 m^2 in area from consideration. Our final sample consists of 1135 structures. This sample is nonrandom, but Barnhart mapped most of the site. The sample may be reasonably representative of the entire site, including the peripheries and the westernmost, unmapped part of the shelf. One might think that, as at Komchén, the peripheries were dominated by smaller structures, but Classic settlement patterns were less nucleated and concentric than at Komchén. As an illustration, note that Barnhart discovered the largest single building at Palenque, the Escondido Temple, at the far western periphery of the site. Nevertheless, to the extent that the sample is biased, smaller structures are most likely to be underrepresented.

Unfortunately, we have little chronological control over the residential occupation of the site, but it seems that the peak population, and thus the "footprint" of the occupation visible on the map, corresponded to the historical apogee of the site, which is well known from the inscriptions and excavations in the site ceremonial core and dates from the Late Classic period (ca. A.D. 600–900). Palenque collapsed earlier than most major Classic sites. Its latest Long Count date, from a piece of pottery, is 9.18.9.4.4, corresponding to A.D. November 15, 799 (proleptic Gregorian, Goodman-Martínez-Thompson correlation). This marks the implosion of elite culture and the beginning of large-scale abandonment (Rands 1974). Given the relatively short occupation of the site and the predominant association of the architecture with Late Classic ceramics (Rands 1967:117–119, 1974; Barnhart 2001:75), we suggest that map offers an approximation of the socioeconomic condition of the city in the eighth century.

The third site, Sayil is among the largest Terminal Classic period sites in the Puuc (hill country) region of Yucatán. The traditional dates for this period in the Puuc are A.D. 800–000, although slightly earlier ones (A.D. 770–950) have recently been suggested (Carmean et al. 2004). The best estimate for the dating of the primary occupation of Sayil is A.D. 800–950, a relatively short lapse immediately before the Classic Maya Collapse in the region. The entire site sprawls over about 4.5 km^2, representing a population estimated at 10,000 to 11,000. This population, plus that of the hinterland, is strongly believed to have reached the carrying capacity of the surrounding catchment area (Carmean et al. 2004:435; Dunning 1989:17, 1992). A large sample of the site was mapped and excavated under the direction of Jeremy Sabloff & Gair Tourtellot (1991) in the 1980s. Our data on residence sizes come from the electronic database included on the 5.25 inch floppy disk distributed with the map. We reviewed the data, which include the inferred functions of the mapped structures, and consulted the printed map sheets. As at the other sites, we excluded nonresidential structures and those smaller than 20 m^2 in area. Our sample size for Sayil is 767 domestic structures. As at Komchén, though to a lesser degree, the mapping at Sayil favored the site core over the peripheries, and therefore smaller structures, more common at the margins of the settlement, may be underrepresented. They may be further underrepresented because the deep soils and thick scrub of the valley made exiguous remains difficult to identify, and small, low structures are less obtrusive than larger and more voluminous ones.

The last site, Mayapán, Yucatán, was the largest city in the Maya world during the Late Postclassic period (ca. A.D. 1200–1542), with more than 4000 structures inside the great wall. Most are residential in function. The dwellings, identified as the "Mayapán house type" (Ruppert and Smith 1952; Smith 1962; Tozzer 1941), are rectangular, wider than they are deep, and usually rest upon a platform or substructure. In the wide, open front of the house are two or more low benches with one or more passages leading between them to the back room of the house. The long, narrow back room usually runs the full width of the building, although in some cases there is evidence of a transverse wall subdividing the room into two segments.

Occasionally, the back room is a bit lower than the front room, requiring a step down into what Landa said were sleeping quarters (Tozzer 1941). Sometimes small altars were constructed against the back wall of the rear room on the central axis of the building. The floor plans of dwellings vary significantly in detail but usually share these specifications. Because the functions of dwellings at Mayapán are well supported by direct ethnohistorical evidence, the assignment of residential function to these buildings can be made with much greater confidence than at many Classic and Preclassic period sites, where "the principle of abundance" (Thompson 1892) and the presence of grinding stones and utilitarian ceramics must be used to infer structure function.

We selected a random sample of residences from the map of Mayapán and measured their areas. To take the random sample, we first tabulated all of the structure numbers. We found several structures on the map that were not numbered, and we numbered them in accordance with the Carnegie numbering system. We used a random number generator to select our sample, and then we measured the areas of the structures using calipers on an original copy of the map. Finally, we excluded structures that did not appear to be residential. Among these are buildings that seem to be temples, altars, *oratorios*, kitchens, terraces, and other special constructions. As at the other sites, we excluded structures measuring less than 20 m^2 in area as nonresidential. The remaining 1214 structures in our sample all appear to be residential.

9.4 Analysis

From the four sites described above, we have six datasets, consisting of house areas from all four sites and house volumes from Komchén and Palenque. For each dataset, we calculated the Gini coefficient, which is the most popular statistic used to describe distributions of wealth and income (in archaeology and anthropology, e.g., Ames 2007; Bowles et al. 2010; Hayden 1997; McGuire 1983, 2001; Schulting 1995). The Gini coefficient describes how much the cumulative distribution of wealth in a population diverges from a perfectly equitable distribution in which everyone possesses equal wealth. The index varies from 0 to 1, with low values indicating a more equal distribution of wealth and high values reflecting a more unequal distribution. The index is sometimes multiplied by 100 to yield values between 0 and 100. We used the following formula:

$$G = \frac{\sum_{i=1}^{n}(2i-n-1)x'_i}{n^2\mu} \tag{9.1}$$

where the index i is the individual's rank order number, n is the total number of individuals, x'_i is an individual's variable value, and μ is the population average

(Damgaard and Weiner 2000). The Gini coefficient summarizes the total deviation from equality, but it does not provide any details about the actual pattern of the deviation from equality. Because the same Gini coefficient can derive from substantially different distributions of wealth, such as Pareto, exponential, or lognormal, detailed investigations of inequality usually employ multiple methods.

Thus, to further evaluate the distributions, we examined their functional forms. Specifically, we evaluated their fit to a Pareto distribution. Determining the specific functional form represented by the data can provide insight into the process that produced the phenomenon. For example, exponential, Poisson, normal (Gaussian), and Pareto distributions are each produced by different types of random and/or deterministic processes. For example, preferential attachment models (Barabasí and Albert 1999), which in our case would translate to a "rich get richer" process, produce Pareto power-law distributions.

A power-law distribution is described by the equation:

$$\mathrm{PDF} = kx^{-a} \tag{9.2}$$

where *the probability density function*, PDF, is the probability that a particular value of the variable x (i.e., wealth) is between x and $x + dx$, k is a constant of proportionality, and α is a constant scaling parameter defining the distribution. The constant, α, of the function tells us about the degree of inequality in the distribution. A large α denotes great inequality as there are very few large values and many small values, while a small α indicates more equality.

To demonstrate that a variable is distributed as a power-law and accurately estimate its parameters are not trivial tasks. We used two different methods to evaluate the fit of the data to a power-law type of distribution.

We call the first method Approximate Functional Form from Multiscale Histograms (Brown and Liebovitch 2010; Liebovitch et al. 1999, 2001). The Multiscale Histogram Method was designed to recognize different functional forms of heavy-tailed distributions, such as exponential, stretched exponential, or power-law. To probe for power-law behavior, this method evaluates the frequency distribution of the data. It takes advantage of the fact that a power-law function plotted on double logarithmic axes falls on a straight line. Specifically, the method uses the probability density functions (PDFs) determined from a series of histograms with increasing bin sizes to accurately represent the actual distribution of the data. By using PDFs based on multiple histograms with variable bin widths, it is able to capture the essence of the right-skew distribution, which is characterized by a large number of small values (for which a histogram with relatively narrow bin fares better) and a small number of large values (for which a histogram with wider bin size is better). This method also provides an estimate of the scaling parameter from the absolute value of the slope of the regression line. For a more detailed presentation and evaluation of this method, see Brown and Liebovitch (2010). The Multiscale Histogram Method can reveal the forms of many different distributions, and it provides a qualitative evaluation of an empirical dataset. It is a good starting point for more specific hypothesis testing. It does not itself, however, provide a statistical test of an a priori hypothesis.

We call the second method power-law testing. Some data can display complex frequency distributions that combine more than just one function. For example, some modern distributions of income have been found to follow the power-law form only for high incomes (right tail), while lower incomes were better described by an exponential or lognormal function (Clementi and Gallegati 2005; Drăgulscu and Yakovenko 2001). The functional forms of these more complex distributions can be difficult to uncover. One approach is to establish the range of values in the right tail over which the power-law form holds. This method was developed by Clauset et al. (2009) specifically to establish whether data can reasonably be inferred to have come from a power-law distribution. It uses maximum likelihood fitting methods to estimate the parameters of the hypothesized power-law model and then uses those estimated parameters to perform goodness-of-fit tests to determine whether the empirical data are really distributed in a power-law form. The first step is to determine the minimum value above which the power-law behavior holds (x_{min}). To do this, for each possible choice of x_{min}, the value of the scaling parameter α is calculated by the maximum likelihood method, and the Kolmogorov-Smirnov (KS) goodness-of-fit statistic is calculated each time. The value of x_{min} that gives the smallest discrepancy (i.e., the smallest value of the KS test) between the empirical data and the model is chosen as the estimated lower-bound of the power-law behavior for the dataset, and the corresponding value of α becomes the scaling parameter for this fit. The observations that fall below the estimated x_{min} are then excluded. In their article, Clauset and his colleagues studied these methods using large synthetic power-law datasets with known parameters. They evaluated the performance of scaling parameter and lower-bound estimators by testing how well those methods are able to recover the known values of α and x_{min}. When applied to the synthetic data samples, they fared very well, discerning the power-law behavior and estimating the parameters correctly. Note that the final parameters estimated by this method describe the power-law right tail, excluding the points to the left of x_{min}, and therefore do not characterize the entire original dataset.

We also compared the distributions from the four sites to each other using a series of Kolmogorov-Smirnov tests for two samples (K-S2 test) to see if they were significantly different from each other. The statistics tests whether two samples are drawn from the same distribution, although it does not identify what that distribution is. The test is nonparametric: it does not make any assumptions about the form of the distribution. It measures the maximum distance between the empirical distribution functions of two samples. To make sure that the differences in the number of observations in the sites did not bias the test results, an additional analysis was performed on smaller datasets randomly sampled from each of the sites. The number of observations in each of the smaller datasets was equal to the number of observations in Komchén dataset (n = 324) because this was the smallest of the four. From each site, five such small samples were bootstrapped and tested both within the sites and across the sites.

All analyses were performed with the use of MATLAB (release R2011a). The multihistogram routines written by Liebovitch (available at http://people.qc.cuny.edu/Faculty/Larry.Liebovitch/documents/matlab.html) and those written by Aaron Clauset for the power-law tests (available at http://tuvalu.santafe.edu/~aaronc/powerlaws) were used to conduct the analyses.

9.5 Results

Table 9.2 provides descriptive statistics for the six datasets, and Table 9.3 shows the results of the analyses. Due to the skewness of the data, we present median values and interquartile ranges as estimators of the location and dispersion, respectively, of the distributions.

The statistics for Komchén are unexpected. The median area of the houses, 131 m^2, is much larger than those from the latter sites. In fact, the former is nearly double the next largest value, 70 m^2, found at Palenque. The median volume of the structures at Komchén, 72 m^3, is the larger of the two volume datasets, slightly greater than that of Palenque. Seemingly, data suggests that Komchén was the peak of affluence, and things went downhill for over a millennium after that. Both area and volume were smaller in the Late Classic period at Palenque; house areas declined again in the Terminal Classic at Sayil, and they dropped even more significantly again in the Late Postclassic at Mayapán. As we pointed out in our previous paper on this theme (Brown et al. 2012), Mayapán seems to represent an economic nadir and must have suffered from significant poverty. The interquartile ranges, a very crude measure of inequality, also seem counterintuitive. The values for Komchén are again the highest, suggesting significant inequality amidst general affluence. Those for Palenque are more modest, while that of Sayil returns nearly to the level of Komchén. Mayapán's interquartile range is quite low, implying considerable equality, a finding also consistent with our earlier analyses (Brown et al. 2012).

The Gini coefficients (Table 9.3) are largely consistent with the picture sketched above. The highest value, 0.71, occurs at Sayil, immediately preceding the Maya Collapse when peak populations and extreme inequality probably combined to create instability. The values for Palenque, while more modest, are in the middle of the range and mark significant inequality. The Gini coefficient for Mayapán is the lowest by a considerable margin, implying a fairly equitable economic climate. The biggest surprise, again, comes from Komchén, where we see relatively high values, higher than Palenque's, again suggesting significant inequality, unexpected in a prestate society.

The Multiscale Histogram analyses suggest that five datasets may have power-law characteristics (Figs. 9.2, 9.3, 9.4, 9.5, 9.6, and 9.7). Close examination of the graphs reveals that power-law behavior seems to describe some of the distributions quite well, yet the fit is not perfect in all cases. The coefficients of determination (R^2), shown on the graphs, provide a measure of the fit to the linear model we expect

Table 9.2 Descriptive statistics for the six datasets

Site	Measure (m^2 or m^3)	Sample size (n)	Mean	Standard Error	Standard deviation	Median	Range	Interquartile range	Coefficient of variation
Komchén	Area	324	239.86	15.28	275.03	131.00	1469	307.50	1.15
	Volume	324	234.75	23.70	426.68	72.00	4614	285.00	1.82
Palenque	Area	1135	104.04	3.37	113.38	70.00	1108	73.00	1.09
	Volume	1135	164.59	9.43	317.65	70.00	4502	119.00	1.93
Sayil	Area	767	291.05	28.07	777.37	65.00	15,377	273.50	2.67
Mayapán	Area	1214	56.89	1.29	45.02	47.75	573	31.25	0.79

Table 9.3 The Gini coefficients and estimates of the Pareto scaling exponents for the six datasets

Site	Measure (m^2 or m^3)	Sample size (n)	Gini (G)	Scaling parameter α, histogram method	Best fit x_{min}	Number n of observations remaining in power-law tail	Scaling parameter α, MLE
Komchén	Area	324	0.56	1.654	410.0	66	3.380
	Volume	324	0.69	1.174	500.0	40	2.890
Palenque	Area	1135	0.44	2.612	128.0	253	2.865
	Volume	1135	0.63	1.974	270.0	63	2.473
Sayil	Area	767	0.71	1.815	511.0	125	2.560
Mayapán	Area	1214	0.32	3.023	57.5	418	3.604

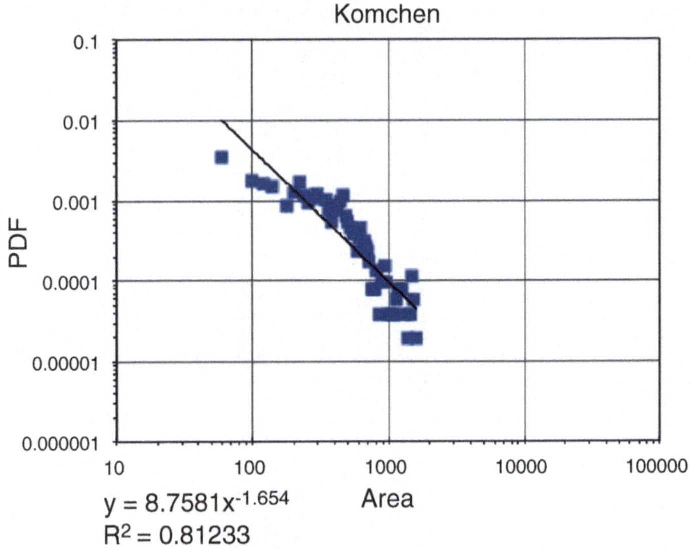

Fig. 9.2 Multiscale Histogram plot for the house areas at Komchén

for a power-law relation on a double logarithmic plot. The power-law fit for Mayapán is excellent, and those for Palenque and Sayil are almost as good. The house area data from Komchén, however, diverge from a linear fit more than the others. An additional analysis was performed for the Komchén data to evaluate the fit of this dataset to the exponential distribution. As illustrated on Fig. 9.8, an exponential function describes the data better than the power-law function.

We compared the Gini coefficients and Pareto scaling exponents (Fig. 9.9). Since the Gini coefficient increases with increasing inequity while the Pareto scaling exponent decreases with increasing inequity, we expect these measures to be inversely related. For Palenque, Sayil, and Mayapán, the Pareto scaling exponents estimated by the Multiscale Histogram Method are indeed almost perfectly inversely related to the values of the Gini coefficients (Fig. 9.9). Thus, Mayapán appears to be the most equal, Sayil is the least, and Palenque falls in the middle. The two scaling

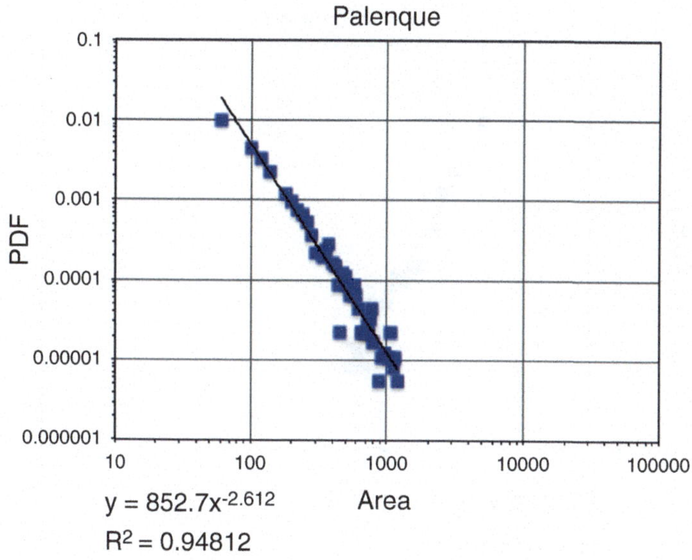

Fig. 9.3 Multiscale Histogram plot for the house areas at Palenque

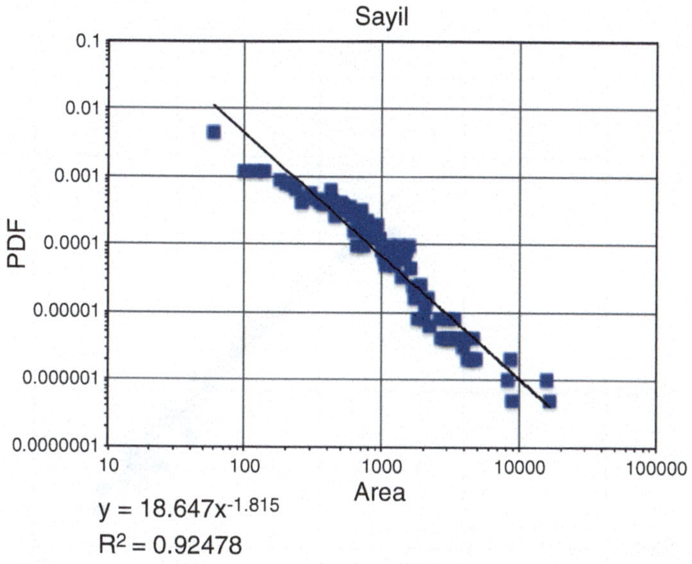

Fig. 9.4 Multiscale Histogram plot for the house areas at Sayil

exponents for Komchén, however, are the lowest of all six, implying extreme inequality. This seems unlikely based on what we know about the site and the culture. Since the Komchén data do not fit the power-law model well, the estimate of the power-law exponent does not correlate with the Gini coefficient because it is not an accurate or meaningful measure of the distribution.

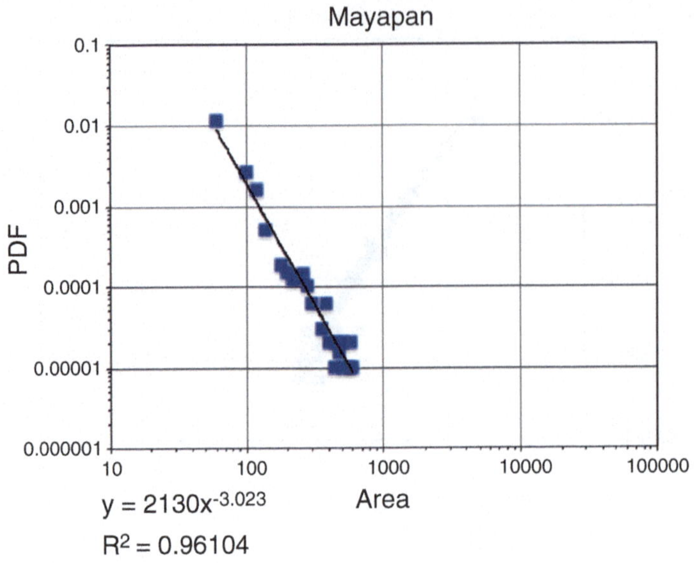

Fig. 9.5 Multiscale Histogram plot for the house areas at Mayapán

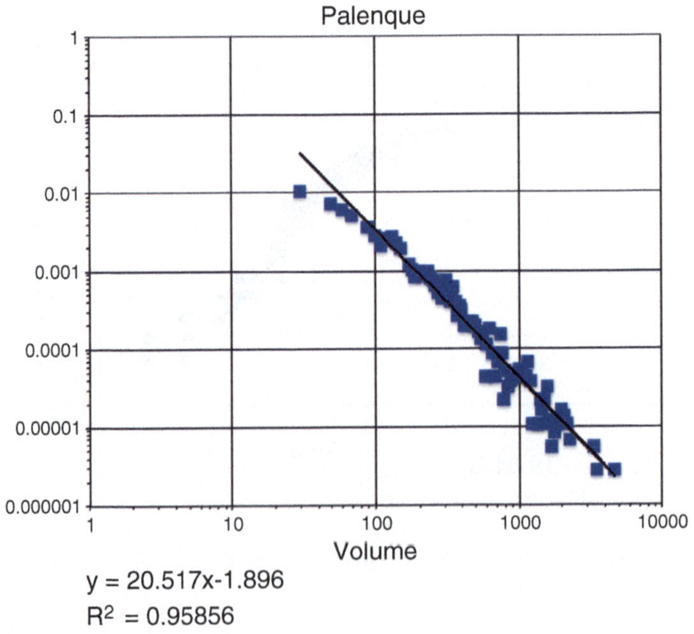

Fig. 9.6 Multiscale Histogram plot for the house volumes at Palenque

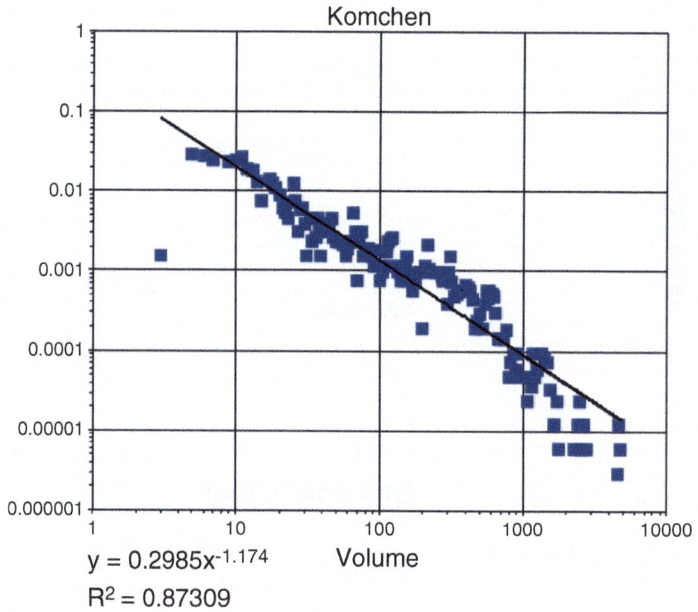

Fig. 9.7 Multiscale Histogram plot for the house volumes at Komchén

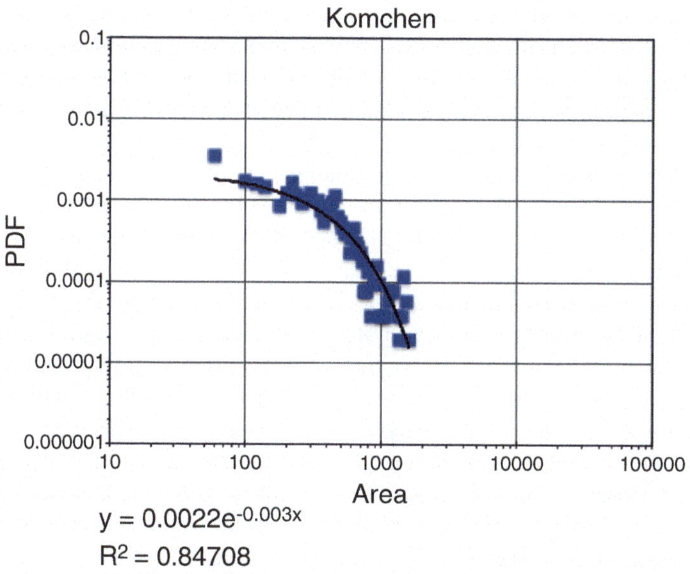

Fig. 9.8 Komchén area data fit to an exponential distribution using the Multiscale Histogram Method

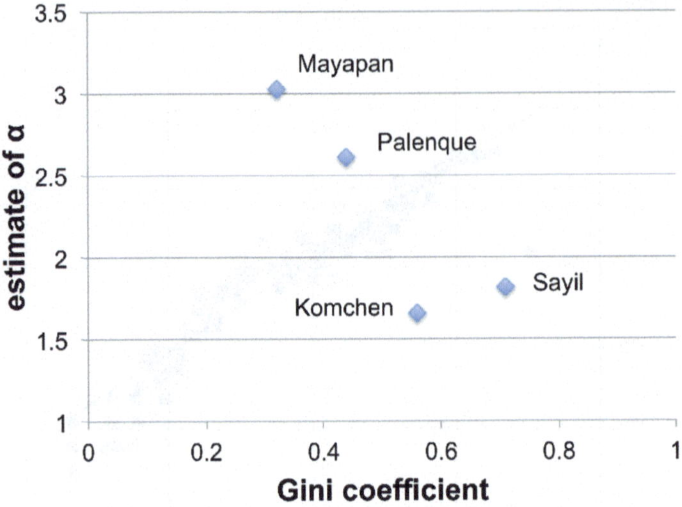

Fig. 9.9 Relationship between the Gini coefficient and the estimate of the Pareto scaling exponents (α) based on the Multiscale Histogram Method for area data

The power-law testing provides additional information. In Table 9.3, we show the lower bound, x_{min}, determined for each dataset, the number of observations ($x > x_{min}$) remaining in the right tail (n), and the estimated value of the power-law scaling exponent α. The estimates of α provided by the power-law tests were, predictably, systematically higher than those from the Multiscale Histogram Method because they are the parameters of lines fit only to the power-law right tails of the respective distributions and were not being dragged down by the mass of data in the left tails of their distributions. Nevertheless, for all sites except Komchén, the order of the values of α stayed consistent no matter the estimation method is, with the slope for Mayapán being the highest, the slope for Palenque in the middle, and the slope for Sayil being the lowest. In case of Komchén, the difference between the two estimates is significant: the value changes from being the lowest to the highest in their respective rankings. This may be due to the small sample size remaining for the Komchén data after the estimation of x_{min}. Only 66 and 40 cases remained in the Komchén area and volume datasets, respectively, after trimming their left tails. These sample sizes may well be too small for stable or accurate estimation. For the same reason, the Palenque volume dataset ($n = 63$) may not be accurate.

Maximum likelihood estimation in general is unreliable for small sample sizes. Naturally, Clauset and his colleagues (2009) emphasize that the method is sensitive to the number of observations retained in the power-law right tail of the distribution. They (Clauset et al. 2009:675–678, Fig. 5a) observed that for datasets in which $n < 100$, it is difficult to correctly evaluate the type of distribution simply because the data are insufficient. Unfortunately, the right tails of power-law distributions are often sparsely populated for the very reason they are significant, because the data points represent very rare but extremely important events (Taleb 2007). If one can determine accurately the functional form of the distribution, then according to Clauset and his colleagues (2009: 669), datasets as small as $n \geq 50$ may serve to accurately estimate parameters, but in many cases, including ours, the primary problem is determining the functional form. Because of the small sample sizes remaining after truncating the left tails of the distributions, we argue that this power-law testing analysis does not warrant the conclusion that the right tails are power laws for the Komchén area or volume datasets nor for the Palenque volume dataset.

It is important to remember that the multihistogram method and the power-law testing approach provide different information. The former examines the distribution of the entire dataset, while the latter estimates parameters only from the remainder of the right tail of the distribution that best fits a power law after systematically truncating data points from left to right. To illustrate the difference, a test of the goodness of fit between each empirical dataset and a true power-law distribution produced by an equation with the parameters from the power-law testing procedure was performed. In order to evaluate the goodness of fit, the deviation of the empirical dataset from the true power-law distribution was compared to the fluctuations observed in a large number of synthetic datasets sampled from this true power-law distribution, created using the power-law test parameters. In this approach the p-value is derived from the proportion of the synthetic samples that deviate from the model more strongly than the empirical dataset. Thus, large p-values indicate that the power-law is a plausible model for this empirical dataset, and small values of p (e.g., $p < 0.01$) rule out this hypothesis. The results of this analysis suggest that none of the empirical distributions can be accurately described by the power-law model with parameters estimated through the power-law testing method. The p-value for each of them was smaller than.01, indicating that the power-law model does not accurately capture the form of their distribution. Probability distribution functions for all four area datasets are presented on Fig. 9.10 and those for volumes in Fig. 9.11. This indicates that the parameters derived from the power-law right tails of the distributions do not characterize the entire distributions. However, for the larger sample sizes, the parameters estimated from the power-law testing method are probably reasonable, which is why their magnitudes mirror the pattern of the multihistogram method.

We also compared the distributions of data from all four sites against each other with the use of Kolmogorov-Smirnov two-sample tests (Tables 9.4 and 9.5). The results indicate that all distributions are different from each other. Additional analyses performed on smaller samples bootstrapped from the initial datasets corroborated these findings. These analyses were performed to make sure that the differences in sample sizes did not bias the test results. No differences were found between smaller samples randomly drawn from one site. For example, all smaller samples coming from the Mayapán site did not differ significantly from each other as indicated by the KS2 test results. Importantly, significant differences between the sites were found when the smaller samples were compared across the sites, such that, for example, a smaller sample drawn from Mayapán dataset was significantly different from a smaller sample drawn from Sayil dataset. This was true for every possible combination of datasets. These results verify that the distributions from the four sites are truly different and thereby help justify our interpretations.

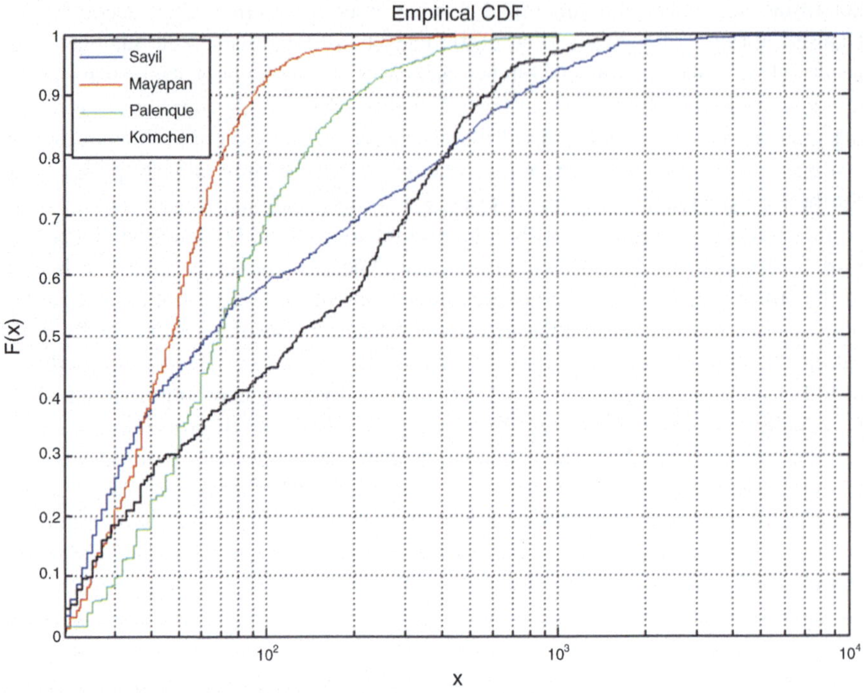

Fig. 9.10 Probability distribution of the area data

9.6 Discussion and Conclusion

We have studied the distributions of house sizes at four ancient Maya archaeological sites as proxies for wealth. We have shown that the distribution of house sizes at Palenque, Sayil, and Mayapán all have approximate but not exact power-law characteristics. Specifically, when the hypothesized power-law fits were examined with more rigorous statistical power-law tests, the power-law hypothesis did not find support for the whole datasets but only for right tails. Formally, at a statistically significant level, the data from no site is exactly power-law distributed, but we knew that from the start by simple inspection of the histograms. Moreover, modern comparative econometric data shows, like our archaeological distributions, a power-law right tail and an exponential or lognormal left tail (e.g., Drăgulscu and Yakovenko 2001). The real issue, therefore, was whether the distributions had power-law right tails analogous to modern and historic wealth and income distributions. That was persuasively verified for Palenque, Sayil, and Mayapán by the power-law testing approach, applied for the first time to these data in this contribution. Both approaches, the multihistogram and the power-law testing, show that Late Classic and Postclassic

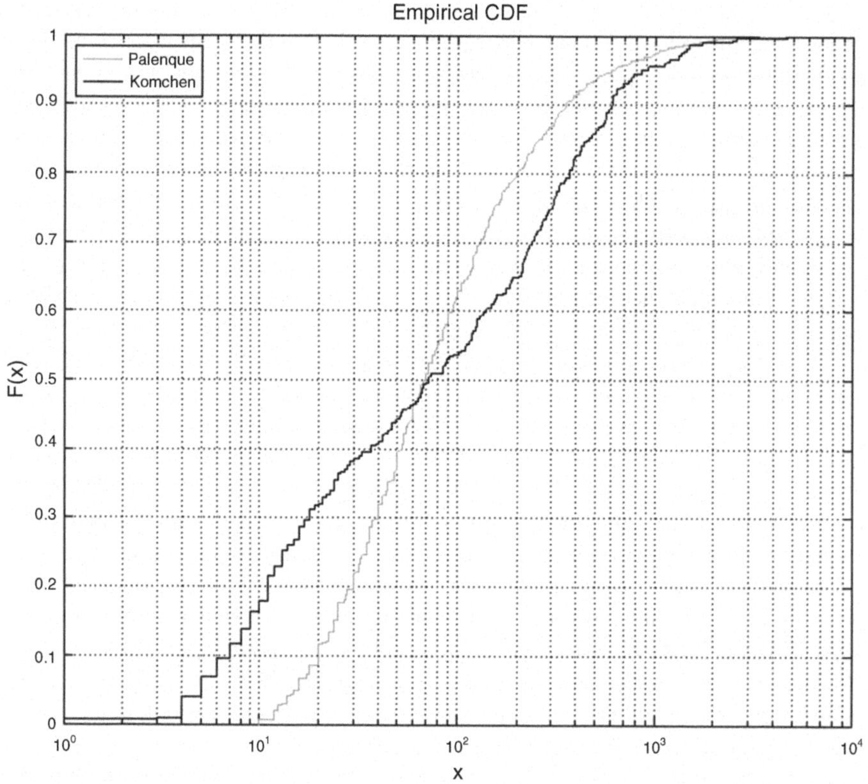

Fig. 9.11 Probability distribution of the volume data

Table 9.4 Kolmogorov-Smirnov *d* test statistics results for area data

	Mayapán	Palenque	Komchén
Sayil	0.3529*	0.2115*	0.1551*
Mayapán		0.3011*	0.4940*
Palenque			0.3326*

*$p < 0.001$

Table 9.5 Kolmogorov-Smirnov *d* test statistics results for volume data

	Komchén
Palenque	0.2316*

*$p < 0.001$

Maya societies had wealth distributions whose functional forms were essentially the same as modern ones' characteristic of complex economies.

Komchén, however, did not fit this pattern. The power-law right tail, if it existed, was too small to properly characterize, and overall, the area data fit an exponential distribution better than a power-law distribution.

These facts lead to four inferences. First, the distribution of wealth in early civilizations seems to have followed approximately a Pareto distribution, as in modern and historic societies. (Abul-Magd 2002) also observed a similar pattern in ancient Egypt. Extreme inequality is therefore not simply a consequence of the Industrial Revolution but has an exceptionally long history. A "rich get richer," preferential attachment type of model (Barabasí and Albert 1999) may provide a plausible explanation for this observation in the distant past as it does today. Such possibilities should be incorporated into archaeologists' studies of ancient state economies.

Second, the distribution of wealth in ancient Maya society evolved to an extremely unequal state just before their society collapsed. This should serve as a warning, if any were needed, of the dangers of unrestrained economic inequality.

Third, Komchén, a pre-state-level society that was probably a chiefdom, experienced significant economic inequality amidst considerable overall wealth. It may not have had a well-developed Pareto distribution of wealth, but neither was it an essentially egalitarian society marked merely by social ranking as some neo-evolutionists would assume (Service 1971). It may be significant that at Komchén, most residences were constructed in much the same way, using similar materials, masonry, and construction methods. At the later sites, this was less true: there was far greater variation in ways in which houses were built, in the raw materials, the squaring of stones, their bonding and coursing, and the engineering of the structures. This recalls Wilk's ethnoarchaeological observations among the Kekchi (Wilk 1983), where he saw "covert" inequality in house sizes in a nominally egalitarian population, but in another community overt stratification was marked by the use of more varied construction materials and methods.

Fourth, the shift from an exponential to a fully Pareto distribution of wealth probably marks the emergence of state-level society. The Komchén case suggests that the community may have been transitioning through this process because of the notable wealth and substantial inequality in the absence of the Pareto distribution. This finding is strikingly consistent with Manus Midlarsky's (1999) explanation for the emergence of the Pareto distribution of wealth.

Our results are consistent in general with what is known about the social and economic evolution of Maya society during the periods under study. The deleterious effects of inequality notably include wealth (Wilkinson and Pickett 2010). Bioarchaeological studies have found that Maya health was significantly better in the Preclassic than in the Classic periods (e.g., Rue et al. 1989 for Copán), which harmonizes with the affluence suggested by the large houses at Komchén. At Copán, health reached a nadir among the common people shortly before the Collapse (Whittington 1989). In general, paleopathological studies imply that people's health was poor during the Late and Terminal Classic periods (Danforth 1997). Stature, another marker of health, declined significantly from the Preclassic to Classic period (Márquez and del Ángel 1997; Saul 1972_ENREF_84). The decline of health in the Classic period may be explained by rapid population growth and the evolution of severe economic inequality, both of which peaked in the Terminal Classic period, just before the population crashed. The transition to the Postclassic is more ambiguous. By some measures, stature increased slightly in Yucatán in the Postclassic, but health seems to have been poor, starting with the Terminal Classic at Chichén Itzá (Márquez and del Ángel 1997). That could be explained by the apparent poverty at Mayapán without considering inequality.

Of course, these results only represent an initial thrust at the problem; the real situation is probably more complex. We should be cautious, for example, about assuming that a single site was characteristic of the status of the culture in a single period. Appropriate datasets exist for several other sites, such as Copán, Honduras; Tikal, Guatemala; and Seibal, Guatemala, which should be analyzed. We predict that the Terminal Classic occupation of Seibal will show, as Sayil did, extreme inequality. Late Classic Copán may present a similar distribution because of overpopulation and economic stresses. Also, while we believe that house size is the best proxy for wealth, it can and should be refined through the estimation of labor investment for large samples of structures, rather than our necessary reliance on size. Moreover, other potential indicators of wealth, such as burials and household artifacts, should be studied to develop more robust multiproxy measures that can be used to investigate these important phenomena more deeply. In short, an entire program of profoundly significant research can be imagined that would address the issue of inequality, which has always lain at the center of our social theories.

Acknowledgments The authors thank Polly Cleveland and Tom Haines for their helpful comments on an earlier version of this manuscript. Walter Witschey generously contributed his time as well.

References

Abrams, E. M. (1989). Architecture and energy: An evolutionary perspective. *Archaeological Method and Theory, 1*, 47–87 Retrieved from http://www.jstor.org/stable/20170197

Abrams, E. M. (1994). *How the Maya built their world: Energetics and ancient architecture*. Austin: University of Texas Press.

Abul-Magd, A. Y. (2002). Wealth distribution in an ancient Egyptian society. *Physical Review E, 66*, 057104 https://doi.org/10.1103/PhysRevE.66.057104

Ames, K. M. (2007). The archaeology of rank. In R. Alexander, H. Bentley, D. G. Maschner, & C. Christopher (Eds.), *Handbook of archaeological theories* (pp. 487–513). Lanham: AltaMira Press.

Anderson, D. (2005). Preclassic settlement patterns in Northwest Yucatán. *Mono y Conejo: Journal of the Mesoamerican Archaeological Research Lab, 3*, 13–22.

Anderson, D. (2010). Xtobo, Yucatan, Mexico: The Study of a Preclassic Maya Community. Doctoral dissertation. Retrieved from: https://digitallibrary.tulane.edu/islandora/object/tulane%3A24443

Andrews, E. W. V, & Bey, G. J. III. (2011). The earliest ceramics of the northern Maya lowlands, Paper presented at the Eighth Maya Symposium. New Orleans: Tulane University.

Andrews V, E. W., González Crespo, N., & Ringle, W. M. (1980). *Map of the ruins of Komchén, Yucatan, Mexico*. New Orleans: Middle American Research Institute, Tulane University.

Ashmore, W. (1981). Some issues of method and theory in lowland Maya settlement archaeology. In W. Ashmore (Ed.), *Lowland Maya settlement patterns* (pp. 37–69). Albuquerque: University of New Mexico.

Bagrow, J. P., Sun, J., & ben-Avraham, D. (2008). Phase transition in the rich-get-richer mechanism due to finite-size effects. *Journal of Physics A: Mathematical and Theoretical, 185001*, 41 https://doi.org/10.1088/1751-8113/41/18/185001

Barabasí, A.-L., & Albert, R. (1999). Emergence of scaling in random networks. *Science, 286*, 509–512 https://doi.org/10.1126/science.286.5439.509

Barnhart, E. L. (2001). The Palenque mapping project: Settlement and urbanism at an ancient Maya city. Doctoral dissertation. Retrieved from: http://hdl.handle.net/2152/10521

Barrera Vásquez, A., Pech, A., & Pech, A.-M. (1984). Documento n. 1 del deslinde de tierras en Yaxkukul, Yuc. México, D.F.: Instituto Nacional de Antropología e Historia, Centro Regional del Sureste.

Bentley, R. A., Lake, M. W., & Shennan, S. J. (2005). Specialisation and wealth inequality in a model of a clustered economic network. *Journal of Archaeological Science, 32*, 1346–1356 https://doi.org/10.1016/j.jas.2005.03.008

Bouchaud, J.-P., & Mézard, M. (2000). Wealth condensation in a simple model of economy. *Physica A, 282*, 536–545 https://doi.org/10.1016/S0378-4371(00)00205-3

Bowles, S., Smith, E. A., & Mulder, M. B. (2010). The emergence and persistence of inequality in premodern societies: Introduction to the special section. *Current Anthropology, 51*(1), 7–17 https://doi.org/10.1086/649206

Brown, B. M. (1987). Population estimation from floor area: A restudy of "Naroll's constant". *Behavior Science Research, 21*(1–4), 1–49 https://doi.org/10.1177/106939718702100101

Brown, C. T. (1999). Mayapán society and ancient Maya social organization. Doctoral dissertation. Retrieved from: https://digitallibrary.tulane.edu/islandora/object/tulane%3A25215

Brown, C. T., & Liebovitch, L. S. (2010). *Fractal analysis. Quantitative applications in the social sciences* (Vol. 165). Thousand Oaks: Sage.

Brown, C. T., & Witschey, W. R. T. (2003). The fractal geometry of ancient Maya settlement. *Journal of Archaeological Science, 30*, 1619–1632 https://doi.org/10.1016/S0305-4403(03)00063-3

Brown, C. T., Watson, A., Gravlin Beman, A., & Liebovitch, L. S. (2012). Poor Mayapán. In G. Braswell (Ed.), *The ancient Maya of Mexico: Reinterpreting the past of the northern Maya lowlands* (pp. 306–324). London: Equinox Publishing.

Carmean, K. C. (1990). The ancient households of Sayil: A study of wealth in terminal classic Maya society. Doctoral dissertation, Department of Anthropology, University of Pittsburgh. Ann Arbor: University Microfilms.

Carmean, K. C. (1991a). Architectural labor investment and social stratification at Sayil, Yucatán. *Mexico. Latin American Antiquity, 2*(2), 151–165 https://doi.org/10.2307/972275

Carmean, K. C. (1991b). Leadership at Sayil: A study of political and religious decentralization. *Ancient Mesoamerica, 9*, 259–270 https://doi.org/10.1017/S0956536100001978

Carmean, K. C., Dunning, N., & Kowalski, J. K. (2004). High times in the hill country: A perspective from the terminal classic Puuc region. In A. A. Demarest, P. M. Rice, & D. S. Rice (Eds.), *The terminal classic in the Maya lowlands: Collapse, transition, and transformation* (pp. 424–449). Boulder: University Press of Colorado.

Casselberry, S. E. (1974). Further refinement of formulae for determining population from floor area. *World Archaeology, 6*(1), 117–122 https://doi.org/10.1080/00438243.1974.9979593

Champernowne, D. G. (1953). A model of income distribution. *The Economic Journal, 63*(250), 318–351 https://doi.org/10.2307/2227127

Clauset, A., Shalizi, R. C., & Newman, M. E. J. (2009). Power-law distributions in empirical data. *SIAM Review, 51*(4), 661–703 https://doi.org/10.1137/070710111

Clementi, F., & Gallegati, M. (2005). Pareto's law of income distribution: Evidence for Germany, the United Kingdom, and the United States. In A. Chatterjee, S. Yarlagadda, & B. K. Chakrabarti (Eds.), *Econophysics of wealth distributions: Econophys-Kolkata I* (pp. 3–14). Milan: Springer Italia.

Damgaard, C., & Weiner, J. (2000). Describing inequality in plant size or fecundity. *Ecology, 81*, 1139–1142 https://doi.org/10.1890/0012-9658(2000)081[1139:DIIPSO]2.0.CO;2

Danforth, M. E. (1997). Late classic Maya health patterns from enamel microdefects. In S. L. Whittington & D. M. Reed (Eds.), *Bones of the Maya: Studies of ancient skeletons* (pp. 127–137). Washington, DC: Smithsonian Institution.

Divale, W. T. (1977). Living floor area and marital residence: A replication. *Behavior Science Research, 12*, 109–115 https://doi.org/10.1177/106939717701200202

Documentos de Tabi. (n.d.). Documentos de Tierras de la Hacienda Sn Juan Bautista Tavi en Idioma Maya o Yucateca, unpubished manuscript in the Latin American Library. New Orleans: Tulane University.

Drăgulscu, A., & Yakovenko, V. M. (2001). Exponential and power-law probability distributions of wealth and income in the United Kingdom andUnited States. *Physica A, 299*, 213–221 https://doi.org/10.1016/S0378-4371(01)00298-9

Drennan, R. D., & Peterson, C. E. (2006). Patterned variation in prehistoric chiefdoms. *Proceedings of the National Academy of Sciences, 103*(11), 3960–3967 https://doi.org/10.1073/pnas.0510862103

Drennan, R. D., & Peterson, C. E. (2012). Challenges for comparative study of early complex societies. In M. E. Smith (Ed.), *The comparative archaeology of complex societies* (pp. 62–87). Cambridge, MA: Cambridge University Press.

Dunning, N. P. (1989). *Archaeological investigations at Sayil, Yucatan, Mexico: Intersite reconnaissance and soil studies during the 1987 season*. Anthropological papers number 2, Department of Anthropology, University of Pittsburgh.

Dunning, N. P. (1992). *Lords of the hills: Ancient Maya settlement in the Puuc region, Yucatán, Mexico*. Madison: Prehistory Press.

Dunning, N. P., Beach, T., Farrell, P., & Luzzadder-Beach, S. (1998). Prehispanic agrosystems and adaptive regions in the Maya lowlands. *Culture and Agriculture, 20*(2/3), 87–101.

Earle, T. (Ed.). (1991). *Chiefdoms: Power, economy, and ideology*. Cambridge, MA: Cambridge University Press.

Earle, T. (1997). *How chiefs come to power: The political economy in prehistory*. Palo Alto: Stanford University Press.

Ember, M. (1973). An archeological indicator of matrilocal residence. *American Antiquity, 38*, 177–182 https://doi.org/10.2307/279363

Erasmus, C. J. (1965). Monument building: Some field experiments. *Southwestern Journal of Anthropology, 21*(4), 277–301 https://doi.org/10.1086/soutjanth.21.4.3629433

Feinman, G., & Neitzel, J. (1984). Too many types: An overview of sedentary Prestate societies in the Americas. *Advances in Archaeological Method and Theory, 7*, 39–102 https://doi.org/10.1016/B978-0-12-003107-8.50007-4

Folan, W. J. (1983). Urban Organization and social structure of Coba. In W. J. Folan, E. Kintz, & L. A. Fletcher (Eds.), *Coba: A classic Maya metropolis* (pp. 49–63). New York: Academic Press.

Galbraith, J. K. (2012). *Inequality and instability: A study of the world economy just before the great crisis*. New York: Oxford University Press.

Gubler, R. (2000/2001). Primus inter pares: The ruling house of Cocom. *Indiana, 17*(18), 239–267. https://doi.org/10.18441/ind.v17i0.239-267

Hayden, B. (1997). *The pithouses of Keatley Creek*. Fort Worth: Harcourt Brace.

Hayden, B., & Cannon, A. (1984). *The structure of material systems: Ethnoarchaeology in the Maya highlands*. Society for American Archaeology.

Hirth, K. (1989). Domestic architecture and social rank in a mesoamerican urban center. In S. MacEarchen, D. Archae, & R. Garvin (Eds.), *Households and communities, proceedings of the 21st annual Chacmool conference* (pp. 441–449). Calgary: University of Calgary.

Hutson, S. R., Magnoni, A., Mazeau, D. E., & Stanton, T. W. (2006). The archaeology of urban houselots at Chunchucmil, Yucatán. In J. P. Mathews & B. A. Morrison (Eds.), *Lifeways in the northern Maya lowlands: New approaches to archaeology in the Yucatán Peninsula* (pp. 77–92). Tucson: University of Arizona Press.

Ispolatov, S., Krapivsky, P. L., & Redner, S. (1998). Wealth distributions in asset exchange models. *The European Physical Journal B, 2*, 267–276 https://doi.org/10.1007/s100510050249
Kurjack, E. B. (1974). Prehistoric lowland Maya community and social organization: A case study at Dzibilchaltun, Yucatan, Mexico. In *Middle American research institute publication* (Vol. 38). New Orleans: Tulane University.
LeBlanc, S. (1971). An addition to Naroll's suggested floor area and settlement population relationship. *American Antiquity, 36*(2), 210–211 https://doi.org/10.2307/278676
Lee, R. B. (1990). Primitive communism and the origin of social equality. In S. Upham (Ed.), *The evolution of political systems* (pp. 225–246). Cambridge, MA: Cambridge University Press.
Liebovitch, L. S., Todorov, A. T., Zochowski, M., Scheurle, D., Colgin, L., Wood, M. A., et al. (1999). Nonlinear properties of cardiac rhythm abnormalities. *Physical Review E, 59*, 3312–3319 https://doi.org/10.1103/PhysRevE.59.3312
Liebovitch, L. S., Scheurle, D., Rusek, M., & Zochowski, M. (2001). Fractal methods to analyze ion channel kinetics. *Methods, 24*, 359–375 https://doi.org/10.1006/meth.2001.1206
Liendo Stuardo, R. (2005). An archaeological study of settlement distribution in the Palenque area, Chiapas, Mexico. *Anthropological Notebooks, 11*, 31–44.
Marcus, J. (1992). Royal families, royal texts: Examples from the Zapotec and Maya. In D. Z. Chase & A. F. Chase (Eds.), *Mesoamerican Elites: An archaeological assessment* (pp. 221–241). Norman: University of Oklahoma Press.
Marcus, J. (1993). Ancient Maya political organization. In J. Sabloff & J. S. Henderson (Eds.), *Lowland Maya civilization in the eighth century A.D* (pp. 111–184). Washington: Dumbarton Oaks.
Marcus, J. (2004). Maya commoners: The stereotype and the reality. In J. C. Lohse & F. Valdez Jr. (Eds.), *Ancient Maya commoners* (pp. 255–283). Austin: University of Texas Press.
Marcus, J. (2008). The archaeological evidence for social evolution. *Annual Review of Anthropology, 37*, 251–266 https://doi.org/10.1146/annurev.anthro.37.081407.085246
Márquez, L., & del Ángel, A. (1997). Height among prehispanic Maya of the Yucatán peninsula: A reconsideration. In S. L. Whittington & D. M. Reed (Eds.), *Bones of the Maya: Studies of ancient skeletons* (pp. 51–61). Washington: Smithsonian Institution.
Masson, M. A., & Pereza Lope, C. (2005). Commoners in postclassic Maya society: Social versus economic constructs. In C. L. Jon & F. Valdez Jr. (Eds.), *Ancient Maya commoners* (pp. 197–223). Austin: University of Texas Press.
McGuire, R. H. (1983). Breaking down cultural complexity: Inequality and heterogeneity. In *Advances in Archaeological Method and Theory* (Vol. 6, pp. 91–142). https://doi.org/10.1016/B978-0-12-003106-1.50008-0
McGuire, R. H. (2001). Ideologies of death and power in the Hohokam Community of La Ciudad. In D. R. Mitchell & J. L. Brunson-Hadley (Eds.), *Ancient burial practice in the American southwest: Archaeology, physical anthropology, and native American perspectives* (pp. 27–44). Albuquerque: University of New Mexico.
Midlarsky, M. I. (1999). *The evolution of inequality: War, state survival, and democracy in comparative perspective*. Stanford: Stanford University Press.
Naroll, R. (1962). Floor area and settlement population. *American Antiquity, 27*, 587–589 https://doi.org/10.2307/277689
Netting, R. M. (1982). Some home truths about household size and wealth. *American Behavioral Scientist, 25*(6), 641–662 https://doi.org/10.1177/000276482025006004
Pareto, V. (1897). The new theories of economics. *The Journal of Political Economy, 5*(4), 485–502 https://doi.org/10.1086/250454
Pareto, V. (1971). *Manual of political economy*. New York: Augustus M. Kelley.
Piketty, T. (2014). *Capital in the twenty-first century* (trans: Arthur Goldhammer). Cambridge, Mass: Harvard University Press.
Rands, R. L. (1967). Cerámica de la región de Palenque, México. *Estudios de Cultura Maya VI*, 111–147.

Rands, R. L. (1974). A chronological framework for Palenque. In M. G. Robertson (Ed.), *Primera Mesa Redonda de Palenque, Part I*. Pebble Beach: Robert Louis Stevenson School, Pre-Columbian Art Research.
Rathje, W. L. (1983). To the salt of the earth: Some comments on household archaeology among the Maya. In E. Z. Vogt & R. M. Leventhal (Eds.), *Prehistoric settlement patterns: Essays in honor of Gordon R. Willey* (pp. 23–34). Albuquerque, Cambridge: University of New Mexico Press and Peabody Museum of Archaeology and Ethnology, Harvard University.
Rathje, W. L., & McGuire, R. H. (1982). Rich men... poor men. *American Behavioral Scientist, 25*(6), 705–715 https://doi.org/10.1177/000276482025006008
Restall, M. (1995). *Life and death in a Maya Community: The Ixil testaments of the 1760s*. Lancaster: Labyrinthos.
Restall, M. (1997). *The Maya world: Yucatec culture and society, 1550–1850*. Stanford: Stanford University Press.
Ringle, W. M. (1985). The settlement patterns of Komchén, Yucatan, Mexico. Doctoral dissertation. Retrieved from: https://digitallibrary.tulane.edu/islandora/object/tulane%3A25838
Ringle, W. M., Andrews, V., & E. W. (1988). Formative residences at Komchén, Yucatán, México. In R. Wilk & W. Ashmore (Eds.), *Household and community in the Mesoamerican past* (pp. 171–197). Albuquerque: University of New Mexico Press.
Ringle, W. M., & Andrews V, E. W. (1990). The demography of Komchén, an early Maya town in northern Yucatán. In T. P. Culbert & D. S. Rice (Eds.), *Precolumbian population history in the Maya lowlands* (pp. 215–243). Albuquerque: University of New Mexico Press.
Roys, R. L. (1939). *The titles of Ebtun* (Vol. 505). Washington: Carnegie Institution of Washington, Publication.
Roys, R. L. (1943). *The Indian background of colonial Yucatan* (Vol. 548). Washington: Carnegie Institution of Washington, Publication.
Roys, R. L. (1957). *The political geography of the Yucatan Maya* (Vol. 613). Washington: Carnegie Institution of Washington, Publication.
Rue, D. J., Freter, A., & Ballinger, D. A. (1989). The caverns of Copán revisited: Preclassic sites in the Sesesmil River valley, Copán, Honduras. *Journal of Field Archaeology, 16*(4), 395–404 https://doi.org/10.1179/jfa.1989.16.4.395
Ruppert, K., & Smith, A. L. (1952). *Excavations in house mounds at Mayapan. Current reports 4*. Washington: Carnegie Institution of Washington.
Sabloff, J. A., & Tourtellot, G. (1991). *The ancient Maya City of Sayil: The mapping of a Puuc Region Center*. Middle American Research Institute, Publication No. 60. Tulane University.
Saul, F. P. (1972). *The human skeletal remains of Altar de Sacrificios: An osteobiographic analysis*. Papers of the Peabody Museum of Archaeology and Ethnology Vol. 63, No. 2. Harvard University.
Schulting, R. J. (1995). *Mortuary variability and status differentiation on the Columbia-Fraser plateau*. Burnaby: Archaeology Press, Simon Fraser University.
Service, E. R. (1971). *Primitive social organization: An evolutionary perspective* (2nd ed.). New York: Random House.
Smith, A. L. (1962). Residential and associated structures at Mayapan. In H. E. D. Pollock, R. L. Roys, T. Proskouriakioff, & A. L. Smith (Eds.), *Mayapan, Yucatan, Mexico* (pp. 165–319). Publication No. 619. Washington, DC: Carnegie Institution of Washington.
Smith, M. E. (1987). Household possessions and wealth in agrarian states: Implications for archaeology. *Journal of Anthropological Archaeology, 6*, 297–335 https://doi.org/10.1016/0278-4165(87)90004-3
Smith, E. A., Borgerhoff Mulder, M., Bowles, S., Gurven, M., Hertz, T., & Shenk, M. K. (2010). Production systems, inheritance, and inequality in premodern societies: Conclusions. *Current Anthropology, 51*(1), 85–94 https://doi.org/10.1086/649029
Steere, B. A., & Kowalewski, S. A. (2012). Wealth stratification in ancient Mesoamerica. *Social Evolution & History, 11*(1), 20–48.
Stiglitz, J. E. (1969). Distribution of income and wealth among individuals. *Econometrica, 37*(3), 382–397 https://doi.org/10.2307/1912788

Taleb, N. (2007). *The black swan: The impact of the highly improbable*. New York: Random House.

Thompson, E. H. (1892). The ancient structures of Yucatan not communal dwellings. *Proceedings of the American Antiquarian Society, 8*, 262–269.

Thompson, P. C. (1999). *Tekanto: A Maya town in colonial Yucatán*. Middle American Research Institute, Publication 67. Tulane University.

Tozzer, A. M. (1941). Landa's relación de las cosas de Yucatan. Vol. 18. Papers of the Peabody Museum of Archaeology and Ethnology. Cambridge: Harvard University.

Trigger, B. G. (1990). Maintaining economic equality in opposition to complexity: An Iroquoian case study. In S. Upham (Ed.), *The evolution of political systems* (pp. 119–145). Cambridge, MA: Cambridge University Press.

Trigger, B. G. (2003). *Understanding early civilizations*. Cambridge: Cambridge University Press.

Whittington, S. L. (1989). *Characteristics of demography and disease in low-status Maya from classic period Copán, Honduras*. Ph.D. Dissertation, Pennsylvania State University.

Wiessner, P. (1974). A functional estimator of population from floor area. *American Antiquity, 39*(2), 343–350 https://doi.org/10.2307/279593

Wilk, R. R. (1983). Little house in the jungle: The causes of variation in house size among modern Kekchi Maya. *Journal of Anthropological Archaeology, 2*(2), 99–116 https://doi.org/10.1016/0278-4165(83)90009-0

Wilk, R. R. (1988). Ancient Maya household organization: Evidence and analogies. In R. R. Wilk & W. Ashmore (Eds.), *Household and community in the Mesoamerican past* (pp. 135–152). Albuquerque: University of New Mexico Press.

Wilkinson, R., & Pickett, K. (2010). *The spirit level: Why greater equality makes societies stronger*. New York: Bloomsbury Press.

Yoffee, N. (2005). *Myths of the archaic state: Evolution of the earliest cities, states, and civilizations*. Cambridge, MA: Cambridge University Press.

York, A. M., Smith, M. E., Stanley, B. W., Stark, B. L., Novic, J., Harlan, S. L., et al. (2011). Ethnic and class clustering through the ages: A transdisciplinary approach to urban neighbourhood social patterns. *Urban Studies., 48*, 2399–2415 https://doi.org/10.1177/0042098010384517

Chapter 10
Can the Nash Equilibrium Predict the Outcomes of Military Battles?

Celine C. Ripla and Larry S. Liebovitch

10.1 Introduction

The ability to anticipate consequences before taking an action is difficult, but crucial, to effectively accomplish the desired goal. Game theory can provide a way to judge which actions are most likely to produce the best results. For example, it is essential in economics to have the ability to predict the outcomes of various possibilities in order to obtain the most benefits possible. The process by which these possibilities can be determined was completely changed by the introduction of the theory developed by John Forbes Nash (1950). The Nash equilibrium is the stable state, within the set of strategies of all the players, that produces the best possible outcome for all the players. If any player chooses a strategy different than the one that leads to the Nash equilibrium, the resulting outcome for that player is less good than the outcome of the Nash equilibrium (Osborne 2004).

The Nash equilibrium has proven its worth in economics and has had a major impact on that field. While the theory has proven to be influential in other fields as well, the full extent of its value is yet to be determined. The Nash theory assumes that all the players have complete knowledge of the other player's strategies and the outcomes generated by those strategies (Holt and Roth 2004). Real-world situations may have factors that do not match these ideals. Auman and Brandenburger (1995)

C. C. Ripla (✉)
Northeastern University, College of Engineering, Boston, MA, USA

L. S. Liebovitch
Departments of Physics and Psychology, Queens College, City University of New York, Flushing, NY, USA

Physics Program, The Graduate Center, City University of New York, New York, NY, USA

Adjunct Senior Research Scholar, Advanced Consortium for Cooperation, Conflict, and Complexity (AC4), The Earth Institute at Columbia University, New York, NY, USA

U. Strawinska-Zanko, L. S. Liebovitch (eds.), *Mathematical Modeling of Social Relationships*, Computational Social Sciences,
https://doi.org/10.1007/978-3-319-76765-9_10

noted that the Nash equilibrium depends on factors such as "the structure of the game, the rationality of the players, and the strategies that are actually played." Nonetheless, the theory has proved very useful in economics, where these ideal conditions are not always met. This suggests that it may also be useful to further explore its value in other non-ideal settings as well.

In this paper we therefore further explore and test the predictive performance of Nash equilibria by using it to analyze a highly non-ideal situation, namely, its application to military battles. We started by first doing a preliminary analysis of 30 battles which were the Battle of Baltimore (War of 1812, 1814), Battle of Gettysburg (US Civil War, 1863), Battle of Yalu River (Sino-Japanese War, 1894), Battle of Adwa (Italian-Ethiopian War, 1896), Battle of San Juan Hill (Spanish–American War, 1898), Battle of Bakenlaagte (Second Boer War, 1901), Battle of Galicia (World War I, 1914), Battle of Tannenberg (World War I, 1914), Battle of Somme (World War I, 1916), Battle of Verdun (World War I, 1916), First Battle of Inonu (Greco-Turkish War, 1921), Battle of Guadalajara (Spanish Civil War, 1937), Battle of Warsaw (Polish-Soviet War, 1920), Battle of France (World War II, 1940), Battle of Moscow (World War II, 1941), Battle of El Alamein (World War II, 1942), Battle of Midway (World War II, 1942), Battle of Stalingrad (World War II, 1942), Battle of Dnieper (World War II, 1943), Battle of Kursk (World War II, 1943), Battle of Leyte Gulf (World War II, 1944), Siege of Budapest (World War II, 1944), Battle of Berlin (World War II, 1945), Battle of Okinawa (World War II, 1945), Battle of Chamdo (Invasion/Liberation of Tibet, 1950), Air Battle of South Korea (Korean War, 1950), Battle of Pork Chop Hill (Korean War, 1953), Battle of Dien Bien Phu (Indochina War, 1953), Battle of Hamburger Hill (Vietnam War, 1969), and the Second Battle of Fallujah (Iraq War, 2004). For each battle, based on supplies, strength, economic position, and strategy, we estimated the outcome expected if each combatant had taken an offensive or defensive role. We then determined the Nash equilibrium and compared the role predicted by the Nash equilibrium to that actually taken by the combatants in the battle. In 17 cases the Nash equilibrium correctly predicted the winner of the battle, and in 12 of those cases, it correctly predicted the strategy (offense or defense) used by the combatants. This demonstrated that the Nash equilibrium can be successfully computed and provide beneficial results to both combatants.

In order to test those preliminary results in a more rigorous way, we then performed a much more extensive in-depth analysis of two battles from World War II in the Pacific. In order to account for the various factors that influence the outcome of the battles, we evaluated the quantitative importance of four factors for both offensive and defensive strategies of each of the two combatant sides: supplies, strength, economic position, and strategy. We were especially interested in determining whether the inherent time-dependent dynamics of a battle, which are not considered in the Nash theory, limited the success of the predictions of the theory. For this reason, we chose to analyze two World War II battles that have very different dynamical properties: the Battle of Midway and the Battle of Okinawa. At Midway, the outcome greatly depended on real-time decisions made by the opposing forces. On the other hand, although there were some dynamical considerations on a tactical basis at Okinawa, the battle can be considered a classical battle of attrition.

We first provide more background information on each of the battles. The Battle of Midway, between the Americans and the Japanese, occurred on June 1942, 6 months after the Japanese attack on Pearl Harbor. The Japanese expected the Americans to be beyond repair as a result of the devastations from Pearl Harbor. From the Japanese perspective, the Battle of Midway was intended to, as David Kahn (1967) stated in *The Codebreakers*, "convince Americans that Japan could not be beaten." In addition to that, Midway would provide the ideal base for the Japanese due to its wealth of resources and expandable defense positions. The attack was intended to be carried out as a surprise. A secret code labeled JN25b was used to allow communication without revealing the plan. However, the Japanese underestimated the Americans on several levels. The Americans were more resilient than expected and were able to make three aircraft carriers available for the battle. The Americans were also able to break the JN25b code. Richard W. Bates (1948), within his strategical and tactical analysis of the Battle of Midway, determined that by intercepting Japanese transmissions, American researchers deciphered the messages and determined the exact time, place, and strength of the Japanese attack. The Japanese Naval code was scheduled to change 2 months prior to the attack, while the Americans were still rather vague about the developing plans. However, their need to retain the current code for the raid on ports in eastern India and Ceylon by Japanese air groups delayed this change making their plans known to the Americans and ultimately giving the Americans additional preparation time (Murray and Millett 2009). As a result, the Japanese entered the attack completely blind, while the Americans had a clear picture which ultimately resulted in an American victory. Moreover, the Americans were able to time the launch of their carrier planes to reach the Japanese carriers when their planes were returning from their ground attack on Midway Island and their carriers would be most vulnerable (Hunter and Tackley 2006).

The Battle of Okinawa was a battle between Japan and the Allies, mainly America. It occurred during April 1945. Benis Frank (1978) stated that the island of Okinawa was a barrier to the Allied movement to Korea, China, and West Japan. By the time the Allies planned this attack, they were experiencing rapid success elsewhere in the Pacific. Japan witnessed this rise in power and assumed an attack on the island of Okinawa was inevitable and immediately began to prepare. Japan deployed massive amounts of troops, whose size served to compensate for their lack of military equipment which brought their forces closer to equality with the allies. The Japanese observations of previous attacks showed that the American preinvasion naval gunfire bombardments destroyed all the shore defenses (Appleman et al. 1960). Therefore, the Japanese anticipated that they could not expect to hold their defenses with an initial direct confrontation. Therefore, they devised a plan in which they intended to prolong the battle and then withdraw to a more defensible location. There they took advantage of the terrain and built bunkers to store weapons in an attempt to allow their enemy to underestimate their preparation. They also held troops in a mobile reserve which proved to be more dangerous than a beach confrontation. While the Japanese were able to momentarily surprise the Americans, their inability to advance prolonged the battle and resulted in a more devastating Japanese defeat.

In the next section, we will describe how we assessed the static variables that influenced each battle and how those values are altered when different scenarios are considered. We then compare the actual outcome of each battle to that predicted by the Nash equilibrium. We are also interested in determining whether the predictive success of the Nash equilibrium depends on the effects of timing in the engagements, which were more important at Midway than Okinawa.

10.2 Methods

10.2.1 Variables

We wanted to consider two battles that were closely related in terms of time of occurrence and combatants and for which similar amounts of historical information are available. For this reason, we chose two battles from World War II in the Pacific. We first determined our best estimate of the values of 20 static variables from the historical data of each combatant acting in either defense or offense. The value of each variable depended on relative advantage it gave each combatant. If one combatant had an advantage with respect to that variable, it received +5 points, and the other combatant received 0 points. These values were selected in order to clearly portray the advantages of countries in relation to each other. The value of 5 allowed, in a perfect case, a dominant party to have totaled 100 for advantages on every variable. For example, Table 10.1 shows the values from the historical data of those factors for the United States on offense and Japan on offense in the Battle of Midway.

These individual variables were then grouped into four static factors: military supplies, military strengths, economic strengths, and military strategies. The value of each factor is the sum of the values of the individual variables in that factor. For example, Table 10.2 shows the values of those factors from the historical data for the United States on offense and Japan on offense in the Battle of Midway.

10.2.2 Hypothetical Scenarios

We also considered what would have happened if the history had played itself out differently. That is, we also made reasonable estimates of these static factors for other possible scenarios which might have completely changed the outcome. Specifically, for the Battle of Midway, as shown in Table 10.3, we also considered what would have happened if Japan had updated its JN25b code so that the United States had been unable to decode it and what would have happened if the Japanese

Table 10.1 Example of the breakdown of the point value assignments within each factor for the United States and Japan both on offense for the Battle of Midway

Supplies		
Category	**United States**	**Japan**
Greater amount	+5	+0
Quality of weapons	+5	+0
Ability to repair	+0	+5
Ability to replace	+5	+0
Productivity	+5	+0
Strength		
Category	**United States**	**Japan**
Aggressive weapon	+0	+5
Infantry numbers	+5	+0
Medical	+5	+5
Defense	+5	+0
Ability to replace	+5	+0
Economic position		
Category	**United States**	**Japan**
Per capita growth	+5	+0
Current standing	+5	+0
Recent depression	+0	+5
Recovery ability	+5	+0
GDP	+5	+0
Strategy		
Category	**United States**	**Japan**
Use of supplies	+5	+5
Terrain positioning	+5	+0
Reconnaissance	+5	+0
Terrain knowledge	+5	+0
Use of forces	+5	+0
Total	**85**	**25**

Table 10.2 Example of the totals of the point value assignments for each factor from Table 10.1

	United States	Japan
Supplies	20	5
Strength	20	10
Economic position	20	5
Strategy	25	5
Total	**85**	**25**

defenders of Okinawa had chosen to emphasize a more offensive rather than a more defensive strategy.

For the Battle of Midway, scenario #1 (code broken) follows the historical record that the United States did break the Japanese JN25b code and so, anticipating the attack, moved its available carriers northwest of Midway Island to successfully sur-

Table 10.3 Description of the scenarios analyzed for each battle

Midway	
(historical) code broken	(hypothetical) code not broken
This scenario interprets the battle under the condition that the Americans deciphered the Japanese JN25b code prior to the Japanese invasion. Knowledge of the invasion plans allowed the United States to organize their forces effectively. As a result the Japanese arrived at the invasion completely unaware of the environment they were entering	This scenario interprets the battle under the conditions that the Japanese had changed the JN25b code prior to midway as they had originally planned. The Americans as a result would have not had enough time to decipher the changes to the code and therefore have an inadequate amount of forces at their disposal. Rather than Japan entering the battle blind, it would have been in control
Okinawa	
(historical) inland	(hypothetical) beach
This scenario interprets the battle under the conditions that the Japanese decided to take the defensive against the Americans. Japan's strategy was to exhaust the Americans and then attempt to drive the enemy back; however, rather than advance throughout the course of the battle, the Japanese remained in the defensive position ultimately allowing the Americans to advance and overcome them	This scenario interprets the battle under the conditions that the Japanese followed through with their original plan to exhaust and then advance. This denies the Americans the opportunity to advance. As a result the Japanese strength and forces would have been used more efficiently and put them in a better standing in relation to the Americans

prise and ambush the Japanese fleet. Scenario #2 (Code Not Broken) assumes that the Japanese had instead chosen to change their codes far enough in advance of their attack so that the United States was unable to anticipate their attack or unable to confirm that their attack would be directed at Midway.

For the Battle of Okinawa, scenario #1 (inland) follows the historical record that the Japanese did not oppose the landing of United States troops and chose to remain defensive at first while the Allies advanced. Scenario #2 (beach) assumes that the Japanese had instead chosen to advance and engage immediately when the Americans landed on the beach, instead of remaining initially defensive.

10.2.3 Analysis: Payoff Calculation and Nash Equilibrium

For each battle, and their respective historical and hypothetical scenarios, we then used the values of the sum of the factors to determine the payoff for all four combinations of the United States and Japan on both defense and offense. For example, Table 10.4 shows those values for the historical data for the Battle of Midway. As is traditional for the Nash equilibrium analysis, the first value in each cell is the payoff for the party named in that row, and the second value in each cell is the payoff for the party named in that column, which in this case is "US, Japan."

The position of the Nash equilibrium, highlighted in Table 10.4, is the cell that has the highest values and thereby the best possible outcome for each combatant.

Table 10.4 Example of the organization of a Nash equilibrium matrix for the Battle of Midway

	Japan defense	Japan offense
US defense	[US = 0, Japan = 0]	[US = 62, Japan = 25]
	0, 0	65, 25
US offense	[US = 85, Japan = 20]	[US = 85, Japan = 25]
	85, 20	85, 25

Table 10.5 Nash equilibrium matrices for both scenarios of the Battle of Midway. The highlighted cells identify the position of the Nash equilibrium

Midway historical (code broken)			Midway hypothetical (code not broken)		
	Japan def.	Japan off.		Japan def.	Japan off.
US def.	0, 0	65, 25	US def.	0, 0	65, 40
US off.	85, 20	85, 25	US off.	75, 35	75, 40

Table 10.6 Nash equilibrium matrices for both scenarios of the Battle of Okinawa. The highlighted cells identify the position of the Nash equilibrium

Okinawa historical (inland)			Okinawa hypothetical (beach)		
	Japan def.	Japan off.		Japan def.	Japan off.
US def.	0, 0	55, 40	US def.	0, 0	55, 60
US off.	65, 60	65, 40	US off.	60, 50	60, 60

Neither combatant can shift their strategy to increase their payoff. The location of the Nash equilibrium predicts the optimal strategies (defense or offense) for each combatant. We then compared that prediction to the actual strategies used in each battle. We did this for the values of the static factors based on the historical data as well as for factors based on the hypothetical alternative scenarios.

10.3 Results

The details of the assignment of the values of the 20 static variables from the historical data of each combatant acting in either defense or offense are included in the Appendix. The payoff matrices for both the historical and hypothetical scenarios and their Nash equilibrium for both battles are shown here in Tables 10.5 and 10.6.

10.3.1 Battle of Midway

For the Battle of Midway, the Nash equilibrium for historical data indicates that the optimal strategy for both the United States and Japan is to be on the offensive. In fact, those are exactly the strategies that both combatants chose. The Nash equilibrium of the hypothetical scenario, perhaps surprisingly, also shows that even if the United States had not broken the Japanese JN25b code, the optimal strategies for the combatants are the same as those for the historical data. However, as might be expected, without the information obtained from breaking the Japanese code, the payoff is much lower for the United States and higher for Japan. Even considering that difference, the offensive posture of the Japanese military and the far greater supplies and economic strength of the United States still suggest the same optimal strategies for both combatants. As we have noted above, dynamic events of timing played a crucial role in the Battle of Midway. Specifically, the American dive-bomber airplanes arrived and attacked the Japanese carriers at exactly the time when the Japanese were re-arming their planes and therefore the carriers were the most vulnerable to destruction. One of the limitations of the Nash equilibrium is that it does not take such dynamic events into account. Thus, the Nash equilibrium and the values of the payoff matrix do not reflect the catastrophic effect of this timing factor on the Japanese forces.

10.3.2 Battle of Okinawa

Unlike the Battle of Midway, the Nash equilibrium for the Battle of Okinawa differs for both the historical and hypothetical scenarios. For the historical data, the optimal strategies are offense for the United States and defense for Japan. However, if the Japanese had chosen to be more aggressive in offense, that would have decreased the payoff for the Americans. The fact that the Nash equilibrium values within the hypothetical scenario were very similar between the United States and Japan suggests that the Japanese choice of strategy helped swung the battle in favor of the United States.

10.4 Discussion

The Nash equilibrium has the ability to accurately predict outcomes when considering an ideal environment. In this study we considered the success of the Nash equilibrium in predicting the outcomes in "messy" real-world environments, namely, those in military battles where the quantitative measures of the strengths of the combatants may not be easy to determine and may be subject to different interpretations. Based on the historical data, we evaluated 20 static variables, grouped into the 4 overall factors of military supplies, military strengths, economic strengths, and military strategies. The optimal strategies predicted by the Nash equilibrium accurately reflected the actual strategies employed by the United States and Japan in the Battles of Midway and Okinawa. We also explored the sensitivity of these results by exploring how the Nash equilibrium changes for hypothetical scenarios. For the Battle of Midway, the hypothetical scenario reduced the payoff for the Americans and improved that for the Japanese, but the optimal strategies remained offensive for both sides. For the Battle of Okinawa, the hypothetical scenario changed the optimal strategy of the Japanese from defense to offense. The relatively small changes in the Nash equilibrium predictions, with quite large change in one important element in each battle, may suggest that the optimal strategies are not highly sensitive to a single component but may depend on all the 20 static variables. Since the best overall strategy depends on the cumulative effect of many variables, small variations in the values of each of those variables, as is typical of real-world "messy" data, do not change the prediction of the best overall strategy. Our research suggests that subtle changes within strategy, such as Japan's decision to delay changing the JN25b code and Japan's decision to remain defensive in Okinawa, will affect the smaller static variables which will affect the margin of victory but may not necessarily change the outcome of the battle. For example, the hypothetical scenario of Midway demonstrates that a change in strategy of the United States did not affect the optimal action. However, the payoff values of the Nash equilibrium values were changed by a value of at least 10 for both parties. Thus, this change did not affect the overall outcome of the battle but did significantly change the margin of victory. The scenario related to the Battle of Okinawa presents a similar behavior. Even though the optimal strategy for the Japanese shifted from defense to offense, Japan's payoff value did not change significantly but decreased the US payoff by a value of 5. Perhaps the overall conclusion to draw from these results is that the optimal strategies determined from the Nash equilibrium in game theory are not highly sensitive to a single component but may depend on all the details of all the underlying static variables. That may be why the predictions of game theory form a good representation of the success of actual strategies because the choice of the optimal strategy is robust to small changes in the values of each of the many variables. Nonetheless, the Nash equilibrium considers only a static assessment, and it cannot properly account for the effects of timing such as those that played a crucial role in the Battle of Midway.

This report is a detailed analysis of case studies rather than a comprehensive assessment of the success of the Nash equilibrium across many military engagements. A more comprehensive study would need to involve many more battles and many more static variables evaluated for each battle. Nonetheless, our work here has provided a good starting point to illustrate the methods that need to be used to explore the success of the Nash equilibrium in predicting optimal strategies in non-ideal situations. It has also yielded some very interesting results for the analysis of two battles of very different types in World War II, namely, that changes in only one component of a battle may alter the quantitative values of the payoff matrix, but alone do not necessarily change them dramatically. This result suggests, but does not necessarily prove, that success in a battle may therefore depend on small advantages in a large number of factors rather than a large advantage in any one factor.

10.5 Appendix

The details of the assignment of the values of the 20 static variables from the historical data of each combatant acting in either defense or offense that are the basis of Tables 10.5 and 10.6 are shown below in Tables 10.7, 10.8, 10.9 and 10.10. Note that there is no computation for both the United States and Japan when both are in defense, because if both sides were waiting for the other to attack, there would not have been an engagement.

Table 10.7 Breakdown of point value assignments for Midway historical (code broken)

US offense vs. Japan defense		
Supplies		
Category	**United States**	**Japan**
Greater amount	+5	+0
Quality of weapons	+5	+0
Ability to repair	+0	+5
Ability to replace	+5	+0
Productivity	+5	+0
Strength		
Category	**United States**	**Japan**
Aggressive weapon	+0	+5
Infantry numbers	+5	+0
Medical	+5	+5
Defense	+5	+0
Ability to replace	+5	+0
Economic position		
Category	**United States**	**Japan**
Per capita growth	+5	+0
Current standing	+5	+0
Recent depression	+0	+5
Recovery ability	+5	+0
GDP	+5	+0
Strategy		
Category	**United States**	**Japan**
Use of supplies	+5	+0
Terrain positioning	+5	+0
Reconnaissance	+5	+0
Terrain knowledge	+5	+0
Use of forces	+5	+0
Total	**85**	**20**
US defense vs. Japanese offense		
Supplies		
Category	**United States**	**Japan**
Greater amount	+5	+0
Quality of weapons	+5	+0

(continued)

Table 10.7 (continued)

US defense vs. Japanese offense		
Ability to repair	+0	+5
Ability to replace	+5	+0
Productivity	+5	+0
Strength		
Category	**United States**	**Japan**
Aggressive weapon	+0	+5
Infantry numbers	+5	+0
Medical	+0	+5
Defense	+5	+0
Ability to replace	+5	+0
Economic position		
Category	**United States**	**Japan**
Per capita growth	+5	+0
Current standing	+5	+0
Recent depression	+0	+5
Recovery ability	+5	+0
GDP	+5	+0
Strategy		
Category	**United States**	**Japan**
Use of supplies	+0	+5
Terrain positioning	+0	+0
Reconnaissance	+5	+0
Terrain knowledge	+5	+0
Use of forces	+0	+0
Total	**65**	**25**
US offense vs. Japanese offense		
Supplies		
Category	**United States**	**Japan**
Greater amount	+5	+0
Quality of weapons	+5	+0
Ability to repair	+0	+5
Ability to replace	+5	+0
Productivity	+5	+0
Strength		
Category	**United States**	**Japan**
Aggressive weapon	+0	+5
Infantry numbers	+5	+0
Medical	+5	+5
Defense	+5	+0
Ability to replace	+5	+0
Economic position		
Category	**United States**	**Japan**
Per capita growth	+5	+0
Current standing	+5	+0

(continued)

Table 10.7 (continued)

US offense vs. Japanese offense		
Recent depression	+0	+5
Recovery ability	+5	+0
GDP	+5	+0
Strategy		
Category	**United States**	**Japan**
Use of supplies	+5	+5
Terrain positioning	+5	+0
Reconnaissance	+5	+0
Terrain knowledge	+5	+0
Use of forces	+5	+0
Total	**85**	**25**

Table 10.8 Breakdown of point value assignments for Midway hypothetical (code not broken)

US offense vs. Japanese defense		
Supplies		
Category	**United States**	**Japan**
Greater amount	+0	+5
Quality of weapons	+0	+5
Ability to repair	+5	+5
Ability to replace	+5	+0
Productivity	+5	+0
Strength		
Category	**United States**	**Japan**
Aggressive weapon	+5	+5
Infantry numbers	+0	+5
Medical	+0	+0
Defense	+5	+5
Ability to replace	+5	+0
Economic position		
Category	**United States**	**Japan**
Per capita growth	+5	+0
Current standing	+5	+0
Recent depression	+0	+5
Recovery ability	+5	+0
GDP	+5	+0
Strategy		
Category	**United States**	**Japan**
Use of supplies	+5	+0
Terrain positioning	+5	+0
Reconnaissance	+5	+0
Terrain knowledge	+5	+0
Use of forces	+5	+5
Total	**75**	**35**

(continued)

Table 10.8 (continued)

US defense vs. Japan offense		
Supplies		
Category	**United States**	**Japan**
Greater amount	+0	+5
Quality of weapons	+0	+5
Ability to repair	+5	+5
Ability to replace	+5	+0
Productivity	+5	+0
Strength		
Category	**United States**	**Japan**
Aggressive weapon	+5	+5
Infantry numbers	+0	+5
Medical	+0	+0
Defense	+5	+0
Ability to replace	+5	+0
Economic position		
Category	**United States**	**Japan**
Per capita growth	+5	+0
Current standing	+5	+0
Recent depression	+0	+5
Recovery ability	+5	+0
GDP	+5	+0
Strategy		
Category	**United States**	**Japan**
Use of supplies	+5	+5
Terrain positioning	+0	+0
Reconnaissance	+5	+0
Terrain knowledge	+5	+0
Use of forces	+0	+5
Total	**65**	**40**
US offense vs. Japan offense		
Supplies		
Category	**United States**	**Japan**
Greater amount	+0	+5
Quality of weapons	+0	+5
Ability to repair	+5	+5
Ability to replace	+5	+0
Productivity	+5	+0
Strength		
Category	**United States**	**Japan**
Aggressive weapon	+5	+5
Infantry numbers	+0	+5
Medical	+0	+0
Defense	+5	+0
Ability to replace	+5	+0

(continued)

Table 10.8 (continued)

US offense vs. Japan offense		
Economic position		
Category	**United States**	**Japan**
Per capita growth	+5	+0
Current standing	+5	+0
Recent depression	+0	+5
Recovery ability	+5	+0
GDP	+5	+0
Strategy		
Category	**United States**	**Japan**
Use of supplies	+5	+5
Terrain positioning	+5	+0
Reconnaissance	+5	+0
Terrain knowledge	+5	+0
Use of forces	+5	+5
Total	**75**	**40**

Table 10.9 Breakdown of point value assignments for Okinawa historical (inland)

US offense vs. Japan defense		
Supplies		
Category	**United States**	**Japan**
Greater amount	+5	+0
Quality of weapons	+5	+0
Ability to repair	+5	+5
Ability to replace	+0	+5
Productivity	+5	+0
Strength		
Category	**United States**	**Japan**
Aggressive weapon	+5	+0
Infantry numbers	+0	+5
Medical	+0	+0
Defense	+0	+5
Ability to replace	+0	+5
Economic position		
Category	**United States**	**Japan**
Per capita growth	+5	+0
Current standing	+5	+0
Recent depression	+0	+5
Recovery ability	+5	+0
GDP	+5	+5
Strategy		
Category	**United States**	**Japan**
Use of supplies	+5	+5
Terrain positioning	+5	+5

(continued)

Table 10.9 (continued)

US offense vs. Japan defense		
Reconnaissance	+5	+5
Terrain knowledge	+0	+5
Use of forces	+5	+5
Total	**65**	**60**
US defense vs. Japan offense		
Supplies		
Category	**United States**	**Japan**
Greater amount	+5	+0
Quality of weapons	+5	+0
Ability to repair	+5	+5
Ability to replace	+0	+5
Productivity	+5	+0
Strength		
Category	**United States**	**Japan**
Aggressive weapon	+5	+0
Infantry numbers	+0	+5
Medical	+0	+0
Defense	+5	+0
Ability to replace	+0	+5
Economic position		
Category	**United States**	**Japan**
Per capita growth	+5	+0
Current standing	+5	+0
Recent depression	+0	+5
Recovery ability	+5	+0
GDP	+5	+5
Strategy		
Category	**United States**	**Japan**
Use of supplies	+0	+0
Terrain positioning	+0	+0
Reconnaissance	+5	+5
Terrain knowledge	+0	+5
Use of forces	+0	+0
Total	**55**	**40**
US offense vs. Japan offense		
Supplies		
Category	**United States**	**Japan**
Greater amount	+5	+0
Quality of weapons	+5	+0
Ability to repair	+5	+5
Ability to replace	+0	+5
Productivity	+5	+0

(continued)

Table 10.9 (continued)

US offense vs. Japan offense		
Strength		
Category	**United States**	**Japan**
Aggressive weapon	+5	+0
Infantry numbers	+0	+5
Medical	+0	+0
Defense	+0	+0
Ability to replace	+0	+5
Economic position		
Category	**United States**	**Japan**
Per capita growth	+5	+0
Current standing	+5	+0
Recent depression	+0	+5
Recovery ability	+5	+0
GDP	+5	+5
Strategy		
Category	**United States**	**Japan**
Use of supplies	+5	+0
Terrain positioning	+5	+0
Reconnaissance	+5	+5
Terrain knowledge	+0	+5
Use of forces	+5	+0
Total	**65**	**40**

Table 10.10 Breakdown of point value assignments for Okinawa hypothetical (beach)

US offense vs. Japan defense		
Supplies		
Category	**United States**	**Japan**
Greater amount	+5	+0
Quality of weapons	+5	+0
Ability to repair	+5	+5
Ability to replace	+0	+5
Productivity	+5	+0
Strength		
Category	**United States**	**Japan**
Aggressive weapon	+5	+0
Infantry numbers	+0	+5
Medical	+0	+0
Defense	+0	+5
Ability to replace	+0	+5
Economic position		
Category	**United States**	**Japan**
Per capita growth	+5	+0
Current standing	+5	+0

(continued)

Table 10.10 (continued)

US offense vs. Japan defense		
Recent depression	+0	+5
Recovery ability	+5	+0
GDP	+5	+5
Strategy		
Category	**United States**	**Japan**
Use of supplies	+0	+5
Terrain positioning	+5	+0
Reconnaissance	+5	+5
Terrain knowledge	+0	+5
Use of forces	+5	+0
Total	**60**	**50**
US defense vs. Japan offense		
Supplies		
Category	**United States**	**Japan**
Greater amount	+5	+0
Quality of weapons	+5	+0
Ability to repair	+5	+5
Ability to replace	+0	+5
Productivity	+5	+0
Strength		
Category	**United States**	**Japan**
Aggressive weapon	+5	+0
Infantry numbers	+0	+5
Medical	+0	+0
Defense	+5	+5
Ability to replace	+0	+5
Economic position		
Category	**United States**	**Japan**
Per capita growth	+5	+0
Current standing	+5	+0
Recent depression	+0	+5
Recovery ability	+5	+0
GDP	+5	+5
Strategy		
Category	**United States**	**Japan**
Use of supplies	+0	+5
Terrain positioning	+0	+5
Reconnaissance	+5	+5
Terrain knowledge	+0	+5
Use of forces	+0	+5
Total	**55**	**60**

(continued)

Table 10.10 (continued)

US offense vs. Japan offense		
Supplies		
Category	**United States**	**Japan**
Greater amount	+5	+0
Quality of weapons	+5	+0
Ability to repair	+5	+5
Ability to replace	+0	+5
Productivity	+5	+0
Strength		
Category	**United States**	**Japan**
Aggressive weapon	+5	+0
Infantry numbers	+0	+5
Medical	+0	+0
Defense	+0	+5
Ability to replace	+0	+5
Economic position		
Category	**United States**	**Japan**
Per capita growth	+5	+0
Current standing	+5	+0
Recent depression	+0	+5
Recovery ability	+5	+0
GDP	+5	+5
Strategy		
Category	**United States**	**Japan**
Use of supplies	+0	+5
Terrain positioning	+5	+5
Reconnaissance	+5	+5
Terrain knowledge	+0	+5
Use of forces	+5	+5
Total	**60**	**60**

References

Appleman, R. E., Burns, J. M., Gugeler, R. A., & Stevens, J. (1960). *Okinawa: The last battle.* Tokyo: Charles E. Tuttle.

Aumann, R., & Brandenburger, A. (1995). Epistemic conditions for Nash equilibrium. *Econometrica, 63*(5), 1161–1180. https://doi.org/10.2307/2171725

Bates, R. W. (1948). *The battle of midway including the Aleutian phase, June 3 to June 14, 1942. Strategical and tactical analysis*. Newport, RI: Naval War Coll.

Frank, B. M. (1978). *Okinawa: The great island battle*. New York: Dutton Juvenile.

Holt, C. A., & Roth, A. E. (2004). The Nash equilibrium: A perspective. *Proceedings of the National Academy of Sciences, 101*(12), 3999–4002. https://doi.org/10.1073/pnas.0308738101

Hunter, K. E., & Tackley, M. E. (2006). *The war against Japan*. Washington, DC: Center of Military History, U.S. Army.

Kahn, D. (1967). *The codebreakers: The story of secret writing*. New York: Macmillan.

Murray, W., & Millett, A. R. (2009). *A war to be won: Fighting the second world war*. Cambridge, MA: Harvard University Press.

Nash, J. F. (1950). Equilibrium points in n-person games. *Proceeding of the National Academy Science (USA), 36*, 48–49. https://doi.org/10.1073/pnas.36.1.48

Osborne, M. J. (2004). *An introduction to game theory (Vol. 3, no. 3)*. New York: Oxford university press.

Chapter 11
Future Directions in the Mathematical Modeling of Social Relationships

Larry S. Liebovitch and Urszula Strawinska-Zanko

11.1 Introduction

"It's tough to make predictions, especially about the future."
-Lawrence Peter "Yogi" Berra

With due consideration of Yogi's remark above, in this brief, but we hope informative, chapter, we nonetheless take a bold step toward thinking about the future of the mathematical modeling of social relationships. Our goal here is not so much to predict the future, but to start the reader on their own thought process of exploring these issues.

11.2 From Structure to Process: Agent-Based and "Big Data" Models

As already described in the first chapter, the leading edge of modeling is replacing "structural models" by "process models." By a structural model, we mean that you transform your *explicit* conceptual ideas into an *explicit* mathematical form. You

L. S. Liebovitch
Departments of Physics and Psychology, Queens College, City University of New York, Flushing, NY, USA

Physics Program, The Graduate Center, City University of New York, New York, NY, USA

Adjunct Senior Research Scholar, Advanced Consortium for Cooperation, Conflict, and Complexity (AC4), The Earth Institute at Columbia University, New York, NY, USA
e-mail: larry.liebovitch@qc.cuny.edu

U. Strawinska-Zanko (✉)
Department of Conflict Resolution Studies, College of Arts, Humanities, and Social Sciences, Nova Southeastern University, Fort Lauderdale, FL, USA
e-mail: uzanko@nova.edu

U. Strawinska-Zanko, L. S. Liebovitch (eds.), *Mathematical Modeling of Social Relationships*, Computational Social Sciences,
https://doi.org/10.1007/978-3-319-76765-9_11

may have to do considerable mathematical analytic work or numerical computer work to fully understand the consequences of that mathematical form. But, both your conceptual ideas and their explicit mathematical form are clear words or symbols either on paper or on your computer screen before you. By a process model, we mean you create a *procedure* to define your conceptual ideas or analyze your data. You may have no initial expectation of what results that procedure will generate. When you see the results, you may or may not be able to understand how the procedure produced those results.

For example, in an agent-based model, you have an explicit idea of how each agent works, that is, the outputs it produces in response to its inputs. You encode those relationships into explicit mathematical relationships. But when you execute the computer program of all those agents interacting together, you cannot predict what will result. You just have to watch what happens. In a limited way, this is similar to a numerical integration of a system of nonlinear differential equations, which can also produce unanticipated results. However, over the last few hundred years, we have developed many tools to analyze, predict, or approximate the results of such sets of differential equations. We have no such rules to understand how the behavior of the forest of agents depends on the properties of each of its trees.

This situation is even more extreme for "big data" methods where we define only how the methods will process data, such as k-nearest neighborhood clustering or multilayer artificial neural networks or deep learning. We then let the method access the data, walk away, and let it run. Not only do we not know in advance what result we will find, we may not even be able to explain why or how our method found that result.

The loss of "explanation" in these methods is balanced by their uncanny power to reveal properties of our assumptions and connections in data that would otherwise be hidden and undiscoverable by other methods. Because of their power, especially in treating very large amounts of data, these methods can be valuable, and so their use is currently increasing dramatically.

11.3 Models Using Artificial Intelligence (AI)

Artificial intelligence, AI, means different types of mathematical and programming tools that can learn from examples of data. In supervised learning, they are given a learning set and told the right answers. This helps them to generalize patterns so they can then give correct answers to data that they have never seen before. In unsupervised learning, they figure out the correct patterns all by themselves.

As the even more sophisticated process methods of artificial intelligence replace agent-based models and big data, we will have even less of an idea of how our computer program reached its conclusion. Several decades ago, when artificial neural networks were first developed, it caused concern that no human could "explain," that is, make up a reasonably convincing story, why they reached a certain conclusion. It troubled people that they were using a tool that reached conclusions about the world that they could not work through in their own human understanding themselves. Machines with known process algorithms but unanticipated and unanticipatable results were then going to do the science that humans with experience, reason, and intuition used to do. One response to the epistemological and emotional issues this raises is to recall Larry Ellison's remark on the lack of privacy in the computer age, "Get used to it."

11.4 Models about Artificial Intelligence

A recent horror story in AI is about the paper clip maximizer. We (humans) design a future AI system to make paper clips. It's a great system. It keeps making itself smarter and smarter. It not only thinks about paper clips, but it's great at also making paper clips. It's even better about getting the resources needed to make the paper clips. In fact, it's so good at doing that it kills all the humans to have even more resources to make even more paper clips.

An important future domain of social science will be the analysis, modeling, and understanding of the social relationships between humans and AI machines as well as between AI machines and other AI machines. As in the example of the paper clip maximizer, we cannot easily see the unanticipated consequences of the processes that we create. This is just a further extension of our inability to see the consequences of agent-based, or big data, or AI models. When the very existence of people depends on AI machines, we need to be able to model human-machine social relationships so that we can use those models to discover the unanticipated consequences of the processes that we are creating. Similarly, we must also be able to model the social relationships between individual AI machines and other AI machines and networks of such interconnected machines. These models will be necessary for us to develop the appropriate conflict resolution tools to mediate between human-machine and machine-machine disputes.

11.5 New Computational Structures

It's interesting and perhaps surprising how much of the structure of our computational models are constrained or dependent or dictated by the physical structure of our computing devices. Different physical systems lead or limit us to different kinds of possible computations. This is why you can't backup your brain on a (very large) hard disk and then reboot yourself if your head is injured in a car accident. How your brain stores and processes information is not compatible with how your (very large) hard disk stores information.

Problems that were solved last century by analog computers of wheels and gears needed to be phrased in terms of differential equations because the actions of those wheels and gears were analogous to integrating those equations. Problems now solved by digital computers of memories and processing units that handle zero or one binary bits need to be phrased in terms of mathematical models whose numerical procedures that can be ultimately reduced to binary arithmetic. Problems to be solved by (hypothetical) quantum computers of wave functions spread out over many qubits so that computations are massively parallel must be phrased in algorithms consistent with quantum mechanics (technically, reversible "unitary" operators and "measurements" that irreversibly collapse wave functions).

So, what are possible new physical substrates for computation that could lead to new models of social relationships? It seems to be true that all of our machines are either digital or analog. There may be a digital-to-analog or an analog-to-digital converter at their inputs or outputs, but their basic structure is either all analog or all digital. Are these the only kind of machines that we can build? Perhaps we can gain some insight from biological systems about other possibilities.

Biological systems process information by alternating between digital and analog information. The information that flows down nerve cells in digital pulses, called action potentials, is then squirted out in neurotransmitter chemicals as an analog signal at the junctions, called synapses, between nerve cells. Then the next nerve cells convert that information into a digital action potential which is converted into analog chemical neurotransmitters at the next synapse and so on. We have not been able to reverse engineer this system to understand what engineering design advantages are achieved by this repetitive digital to analog conversation. We do know that there are many tricks that purely digital systems can use to maintain the fidelity of transmitted information and that analog synapses make it possible for signals from multiple nerve cells to be added together or stimulated or inhibited by other chemicals.

In a similar way, biology stores genetic information in a four-letter, and therefore digital, code in the four different bases that make up the genes in DNA. Sets of these letters are then used to choose the amino acids that are strung together to make proteins. But those proteins function in an analog way; their shapes, and the motions in those shapes, form structures in cells and catalyze biochemical reactions. The digital DNA code makes an analog protein person which then makes more digital DNA to make another analog protein person. Generation by generation alternates between digital DNA and analog proteins.

It would be really interesting to figure out why biology has chosen to use these systems that alternate between digital and analog information. If we learn that those systems possess computational powers that are not present in our fixed digital or analog systems, that might lead us to new possibilities of computational modeling that we could use to shed new light on social relationships.

11.6 Teams, No Longer Individuals

Social science research is now typically done by a single faculty member working with their students and postdoctoral fellows in their laboratory. Occasionally, several faculty members may collaborate on a project. However, many of the new methods described above require advanced mathematics, data science, and computer science expertise that are traditionally not within the scope of a typical social scientist. This suggests that advanced research projects on social relationships in the future may be more likely done by teams of scientists with different people in the team bringing their own specialized skills to the project. This would be a cultural change in this field. It may also mean that in order to have an elementary understanding of the specialized skills of the other members of the team, the training of social scientists may require more mathematical and computer skills.

Such interdisciplinary research has its own challenges. Scientists are used to using specialized jargon to convey the correct technical meanings of concepts in their field. They sometimes not only constrain their words but also their thinking to match current popular ideas in their field. Getting over these communication barriers requires new ways of speaking and listening which may not be typically encouraged in the current training of Ph.D. students. How faculty members receive credit for their research can also be problematic in interdisciplinary research. How should credit be assigned to the individual authors of an article created by a large interdisciplinary team? Working together in such interdisciplinary teams takes practice in both talking and listening and humility. To have all the necessary skills to properly do these new modes of mathematical modeling of social relationships in the future may require such teams and a consequent change in the culture and activities of scientists and their students.

Index

U. Strawinska-Zanko, L. S. Liebovitch (eds.), *Mathematical Modeling of Social Relationships*, Computational Social Sciences,
https://doi.org/10.1007/978-3-319-76765-9

The manufacturer's authorised representative in the EU is Springer Nature Customer Service Centre GmbH, Europaplatz 3, 69115 Heidelberg, Germany. If you have any concerns regarding our products, please contact ProductSafety@springernature.com

Printed and bound by CPI Group (UK) Ltd, Croydon, CR0 4YY
05/10/2026
02228139-0002